Lecture Notes in Electrical Engineering

Volume 647

The book series *Lecture Notes in Electrical Engineering* (LNEE) publishes the latest developments in Electrical Engineering—quickly, informally and in high quality. While original research reported in proceedings and monographs has traditionally formed the core of LNEE, we also encourage authors to submit books devoted to supporting student education and professional training in the various fields and applications areas of electrical engineering. The series cover classical and emerging topics concerning:

- Communication Engineering, Information Theory and Networks
- Electronics Engineering and Microelectronics
- Signal, Image and Speech Processing
- Wireless and Mobile Communication
- Circuits and Systems
- Energy Systems, Power Electronics and Electrical Machines
- Electro-optical Engineering
- Instrumentation Engineering
- Avionics Engineering
- Control Systems
- Internet-of-Things and Cybersecurity
- Biomedical Devices, MEMS and NEMS

For general information about this book series, comments or suggestions, please contact leontina.dicecco@springer.com.

To submit a proposal or request further information, please contact the Publishing Editor in your country:

China

Jasmine Dou, Associate Editor (jasmine.dou@springer.com)

India, Japan, Rest of Asia

Swati Meherishi, Executive Editor (Swati.Meherishi@springer.com)

Southeast Asia, Australia, New Zealand

Ramesh Nath Premnath, Editor (ramesh.premnath@springernature.com)

USA, Canada:

Michael Luby, Senior Editor (michael.luby@springer.com)

All other Countries:

Leontina Di Cecco, Senior Editor (leontina.dicecco@springer.com)

**** Indexing: The books of this series are submitted to ISI Proceedings, EI-Compendex, SCOPUS, MetaPress, Web of Science and Springerlink ****

More information about this series at http://www.springer.com/series/7818

Ferran Martín · Cristian Herrojo ·
Javier Mata-Contreras · Ferran Paredes

Time-Domain Signature Barcodes for Chipless-RFID and Sensing Applications

 Springer

Ferran Martín
Departament d'Enginyeria Electrònica
Universitat Autònoma de Barcelona
Bellaterra, Spain

Cristian Herrojo
Departament d'Enginyeria Electrònica
Universitat Autònoma de Barcelona
Bellaterra, Spain

Javier Mata-Contreras
Departamento de Ingeniería de
Comunicaciones
Universidad de Málaga
Málaga, Spain

Ferran Paredes
Departament d'Enginyeria Electrònica
Universitat Autònoma de Barcelona
Bellaterra, Spain

ISSN 1876-1100   ISSN 1876-1119   (electronic)
Lecture Notes in Electrical Engineering
ISBN 978-3-030-39728-9   ISBN 978-3-030-39726-5   (eBook)
https://doi.org/10.1007/978-3-030-39726-5

This Springer imprint is published by the registered company Springer Nature Switzerland AG
The registered company address is: Gewerbestrasse 11, 6330 Cham, Switzerland

To our families
Anna, Alba, and Arnau
Noe and Chloe
Maria, Roman, and Roger
Ana
and to many other people who have
unconditionally, and sometimes
imperceptibly, provided to us their strong
and continuous support

Preface

Automatic identification and data capture (AIDC) systems have progressively increased their presence in our daily lives in recent years. Among them, radiofrequency identification (RFID) is probably the technology that has experienced a most significant growth and penetration in the market. RFID, a wireless technology that uses RF or microwaves for identification and tracking, cannot compete in terms of cost against optical identification systems. For this reason, the AIDC market is still dominated by optical barcodes and quick response (QR) codes. However, RFID exhibits advantages over optical systems, including higher data capacity, larger read ranges, and the possibility to read multiple tags simultaneously, among others. With these advantageous aspects, RFID systems are appropriate in a wide range of markets, where the cost of the tags can be sacrificed in favor of the high performance that RFID systems exhibit. Applications include automated data collection, identification and tracking of items, animals and persons, livestock identification, automated vehicle identification, wireless sensing, etc.

In most commercial RFID systems, the tags are equipped with an integrated circuit (IC), or chip (where data is stored), and with a planar antenna, necessary for communication with the reader. Such integrated circuits represent a huge portion of the cost of chip-based RFID tags and prevent from the further penetration of RFID technology in the market. For instance, the price of passive UHF-RFID tags (fed by the interrogation signal rather than by batteries) is in the range of several eurocents. This cost is small, but it may be prohibitive in many low-cost items.

Chipless-RFID technology emerged several years ago as a low-cost alternative to chipped-RFID, where the tag IC is replaced with a planar (printed or etched) encoder, with substantially lower cost. Despite the intensive research efforts carried out in the last fifteen years, the performance of state-of-the-art chipless-RFID systems, particularly the data storage capacity, is not comparable to the one of the chip-based systems. Most reported chipless-RFID systems store the information in the frequency domain. In such systems, the encoders consist of a set of printed, or etched, resonant elements tuned to different frequencies, and the identification (ID) code is given by the so-called spectral signature (with singularities in the amplitude or phase response of the tag at certain frequencies, depending on the specific ID code). The number of bits of spectral signature barcodes (the usual designation of frequency-domain tags) is limited by the required spectral bandwidth of the interrogation signal necessary for tag reading (proportional to the number of bits). Although several proposals of chipless-RFID systems where the information is stored in frequency and other domains simultaneously have been reported (hybrid systems), it is not possible to compete against the number of bits of chip-based tags for interrogation signals with reasonable bandwidths (increasing the bandwidth severely affects reader cost).

Time-domain chipless-RFID tags have been also reported. Most time-domain systems are based on time-domain reflectometry (TDR), where the ID code is inferred from the echoes of a narrow pulsed (wideband) signal generated by a delay line with reflectors. Commercial TDR systems based on surface acoustic wave (SAW) encoders exhibiting very high data capacity are indeed available. However, the cost of these SAW-based tags is very high. Low-cost TDR-based tags based on printed or etched delay lines of various types have been also reported, but extremely long delay lines or very narrow pulses are needed in order to implement TDR systems able to read tags with a moderate number of bits (still far from the data capacity of chipped tags).

In this book, an unconventional time-domain approach for chipless-RFID that combines low tag cost and high data storage capacity is reported. The tags consist of a set of identical metallic inclusions (the encoder) printed or etched on a dielectric substrate and arranged as a linear (or circular) chain. Tag reading in these systems proceeds by proximity, through near field coupling with the reader, and the bits are read sequentially (by displacing the tag over the reader) in a time-division multiplexing scheme. The reader contains a sensitive element, a resonator-loaded transmission line, able to detect the functional inclusions of the chain, thereby providing the tag ID code. For that purpose, the reader line is fed by a harmonic (single tone) interrogation signal, which is amplitude modulated (AM) at the output port of the line due to tag motion. Specifically, the functional inclusions, usually associated with the logic state '1', modify the transmission coefficient of the reader line, and consequently, the harmonic interrogation signal is AM modulated at the output port of the line. Since the ID code is contained in the envelope function of such AM signal and revealed as singularities (either peaks or dips) in time domain, it follows that such encoders can be designated as *time-domain signature barcodes*. This is the nomenclature adopted in this book for these time-domain chipless tags, in parallelism to the so-called spectral (or frequency domain) signature barcodes, where the ID code is contained in the frequency domain.

The book is organized as follows. In Chap. 1, the advantages and limitations of chipless-RFID versus chip-based RFID and optical barcodes are discussed, and the classification, as well as a comparative analysis, of the state-of-the-art chipless-RFID approaches is reported. Such analysis is mainly done based on the reported data capacity and density of the different systems. Despite the fact that the analysis does not include the time-domain signature barcodes subject of the present work (and left for future chapters), Chap. 1 concludes by pointing out the high achievable number of bits of these encoders, competitive against the data storage capacity of chipped tags (yet keeping relative small tag sizes).

In Chap. 2, the working principle of chipless-RFID systems based on time-domain signature barcodes is presented in detail. The advantages and limitations as compared to other chipless-RFID approaches are presented, with special emphasis on the data density and capacity (the key advantageous aspects of time-domain signature barcodes) and read range. Because tag reading proceeds through near field, proximity and proper alignment between the tag and the reader are required. Although this is a clear drawback in certain applications, where far-field tags are a due, reading by proximity provides a high level of confidence against eavesdropping or spying, and this may be of interest in other applications, e.g., secure paper or authentication, as discussed in Chap. 2. This chapter also discusses strategies to enhance the data density and provides several examples of chipless-RFID systems based on time-domain signature barcodes, including approaches based on encoders implemented by means of rectangular-shaped resonators and linear metallic inclusions. An exhaustive tolerance analysis, to study the effects of potential misalignments between the tag and the reader, as well as variations in their relative distance (e.g., caused by vibrations), is also included (for each specific approach) in the chapter.

System requirements for industrial scenarios and applications are the subject of Chap. 3. Tag programming and erasing, tag implementation in flexible substrates, including plastic and paper, and synchronous tag reading are discussed in this chapter. As it is highlighted, tag programming is a key factor to reduce tag fabrication costs, since thousands or even millions of all-identical tags can be massively manufactured through industrial processes (e.g., rotogravure, screen-printing, offset), and then programmed (i.e., equipped with the ID code) in a later stage (e.g., by means of low-cost process, such as inkjet printing). Since massive tag fabrication involves a single mask, it follows that the reported approach (where tag personalization is separated from the main industrial fabrication process) offers a low-cost solution for tag manufacturing, as reported in Chap. 3. Secure paper, proximity sensors, and motion control are the main relevant applications of the proposed chipless-RFID system. All these applications are included in Chap. 3.

In Chap. 4, a particular application of the reported time-domain signature barcodes is considered: electromagnetic rotary encoders. Such encoders are indeed angular displacement and velocity sensors and are therefore motion control systems. However, rotary encoders based on metallic inclusions deserve a specific chapter, since such encoders may represent an alternative to the well-established optical encoders in applications where cost and operation in harsh and hostile environments are a due. Strategies to enhance the number of pulses per revolution of the envelope function, the figure of merit in rotary encoders (with direct impact on angular resolution), are discussed in the chapter.

Finally, concluding remarks and future prospects are included in Chap. 5. Tag implementation with organic conductive inks (environmentally friendly and fully recyclable) and all-dielectric 3D-printed tags based on permittivity contrast (robust against mechanical wearing) may constitute two topics of research interest for near future, as revealed by the promising preliminary results already obtained.

This book reflects part of the recent research activity carried out by the authors within the topic of chipless-RFID (although in the review Chap. 1, the state of the art in chipless-RFID technology contributed by many authors worldwide is also included). It is the authors' hope that the present manuscript constitutes a significant contribution in the field of RFID and related topics and that the book can be of practical interest to engineers, professionals, and students working in this fascinating sub-area of wireless technologies.

Barcelona, Spain
August 2019
Ferran Martín
Cristian Herrojo
Ferran Paredes
Javier Mata-Contreras

Acknowledgements

The contents of this book are the result of the research activity on chipless-RFID carried out at CIMITEC, Universitat Autònoma de Barcelona, during the last five years. The authors of the manuscript have mainly developed such activity, but many other people have been directly or indirectly involved in this research topic at CIMITEC and have contributed to its success.

In particular, the authors would like to express their most sincere gratitude to Dr. Jordi Naqui, who obtained the Ph.D. degree under the supervision of Prof. Ferran Martín, and who formerly contributed the research topic of this book through preliminary results on electromagnetic rotary encoders. The authors are also in debt to Dr. Gerard Sisó and to Javier Hellín for the fabrication of the different reader prototypes (sensitive part) and the tags implemented on microwave substrates (involving either milling or etching processes). Inkjet-printed tags on plastic or paper substrates have been fabricated at the Institut de Microelectrònica de Barcelona, Centro Nacional de Microelectrònica (CNM-CSIC), and the authors would like to acknowledge Dr. Eloi Ramon, Alba Nuñez, and Dr. Miquel Moras for their effort in providing high-quality tag prototypes. It is also worth mentioning the contribution of these researchers in developing the readout electronics of some prototype readers. Finally, from the technical viewpoint, the authors would like to express their gratitude to Prof. Jordi Bonache and to Álvaro Jaque, who tuned up the experimental setup for the fabrication of the 3D-printed permittivity contrast encoders (presented in Chap. 5 as preliminary results for the future development).

Several funding institutions, agencies, and companies have supported the research activity reflected in this book. Special thanks are given to the Ministry of Economy and Competitiveness and National Research Agency, Spain, and AEI/FEDER European Union for the financial support through project contracts TEC2013-40600-R, TEC2016-75650-R, RTC-2014-2550-7, and RTC-2017-6303-7, to the AGAUR Research Agency, Catalonia Government, for the projects 2014SGR-157, 2017SGR-1159, and 2014LLAV00046, and to the European Space Agency (for project contract 4000111799/14/NL/SC). The following companies have also directly or indirectly supported the research activities carried out by the authors: EMXYS, Scytl, Meditecnologia, Nabelia, and Hohner Automáticos. The authors are also in debt to ACCIÓ (Government of Catalonia) for the continuous support to CIMITEC through the TECNIO network. It is also worth mentioning the support of ICREA (Institució Catalana de Recerca i Estudis Avançats) to Ferran Martín through three ICREA Academia distinctions (calls 2008, 2013, and 2018).

From the point of view of research management, the authors would like to express their gratitude to Gerard Sisó, Executive Manager of CIMITEC, to Anna Cedenilla (also in charge of copyright issues and permissions), and to Marta Mora and her team (the Administrative staff of the Electronics Engineering Department at Universitat Autònoma de Barcelona). We would also like to include in the acknowledgment list the consulting company ASET Solutions, headed by Joan Plana, for his continuous support in creating research consortia and collaborative projects with companies and institutions.

Last but not least, the authors would like to express their warmest gratitude to their respective families for their patience and continuous support during the preparation of this manuscript and during the preceding research activities.

Contents

Acronyms

AIDC	Automatic identification and data capture
AM	Amplitude modulation
BC-SRR	Broadside-coupled split-ring resonator
BC-S-SRR	Broadside-coupled S-shaped split-ring resonator
BER	Bit-error rate
CPW	Coplanar waveguide
CRLH	Composite right-left handed
CSR	Complementary spiral resonator
CSRR	Complementary split-ring resonator
DPF	Density per frequency
DPL	Density per length
DPS	Density per surface
EC-SRR	Edge-coupled split-ring resonator
ED	Envelope detector
ELC	Electric LC
EMI	Electromagnetic interference
EPC	Electronic product code
HF	High frequency
IC	Integrated circuit
ID	Identification
LF	Low frequency
MIW	Magneto-inductive wave
OCR	Optical character recognition
OOK	ON–OFF keying
PCB	Printed circuit board
PEN	Polyethylene naphthalate
PPM	Pulse-position modulation
PPR	Pulses per revolution
QR	Quick response
RCS	Radar cross section

REF	Reference
RF	Radio frequency
RFID	Radio frequency identification
SAW	Surface acoustic wave
SHF	Super high frequency
SMA	Sub-miniature version A
SMD	Surface-mount device
SRR	Split-ring resonator
S-SRR	S-shaped split-ring resonator
TDR	Time-domain reflectometry
UHF	Ultra-high frequency
USB	Universal serial bus
UWB	Ultra-wide band
VCO	Voltage-controlled oscillator

Chapter 1
State-of-the-Art in Chipless-RFID Technology

An unconventional approach for the implementation of chipless-RFID systems and related sensors is reported in this book. As it will be shown in Chap. 2, the tags consist of chains of identical resonant elements or inclusions (either functional or inoperative), etched or printed at predefined positions on a dielectric substrate, and tag reading proceeds by time-division multiplexing through near field. In a reading operation, the tag should be mechanically guided above the sensitive part of the reader, a planar microwave structure able to detect (through near-field coupling and sequentially) the functional and inoperative resonators, or inclusions, of the tag, consequently providing the ID code. To properly understand the advantages and limitations of this chipless-RFID approach, based on time-domain signature barcodes (the designation given to the tags), it is necessary to review the state-of-the-art in chipless-RFID technology. This is the main aim of the present chapter. Nevertheless, a comparative analysis of the main automatic identification systems, including chipless-RFID, chip-based-RFID and optical systems (optical barcodes and QR codes), is also included in this chapter, in order to contextualize the scope of chipless-RFID technology.

1.1 Chipless-RFID Versus Chipped-RFID Technology and Optical Barcodes

Currently available methods to identify and collect data of objects are usually referred to as automatic identification and data capture (AIDC) systems. Optical barcodes, RFID, biometrics (e.g., iris and facial recognition systems), optical character recognition (OCR), smart cards, or voice recognition are various technologies that can be categorized within AIDC systems. Among them, optical barcodes (including also two-dimensional, or QR, codes) have dominated (and still dominate) the AIDC market, mainly due to the marginal cost of the identification labels (a printed pattern), but also related to the robustness of optical systems against reading errors.

© Springer Nature Switzerland AG 2020
F. Martín et al., *Time-Domain Signature Barcodes for Chipless-RFID and Sensing Applications*, Lecture Notes in Electrical Engineering 647,
https://doi.org/10.1007/978-3-030-39726-5_1

Concerning penetration in the market, the second position is occupied by RFID technology. RFID is a wireless AIDC technology that uses radiofrequency waves (including also microwaves or even millimeter waves) for object identification and tracking [1–3]. Most commercially available RFID systems use identification tags equipped with silicon integrated circuits (IC), or chips, which contain the information relative to the tagged item. The system is also composed by the reader, which communicates with the tag through a radio link (or through inductive coupling in near-field systems), and by the middleware, necessary for data collection, processing, and storage (Fig. 1.1). As compared to optical barcodes or QR codes, RFID tags are able to store significantly much more information, being possible the identification of individual items. Moreover, far-field passive RFID tags do not require direct line-of-sight with the reader. Indeed, reading distances of several meters are possible in passive tags operating in the ultra-high frequency (UHF) band, the most extended one, or in the super-high frequency (SHF) band (Table 1.1), and such distances can be extended up to dozens of meters in tags equipped with batteries (active tags), at the expense of higher tag cost and size.

Despite the fact that passive UHF-RFID tags are relatively cheap (see Table 1.1), its use is prohibitive in many applications involving low-cost tagged items. Tag cost is dictated by the presence and placement of the IC (by contrast tag size is determined by the antenna, needed for communication purposes with the reader), and it is difficult to envision a future scenario (at least at short term) where the prices of chipped tags drop to 1 eurocent or below. Within this context, chipless-RFID has emerged as an alternative technology to chip-based RFID to partially alleviate the high cost of

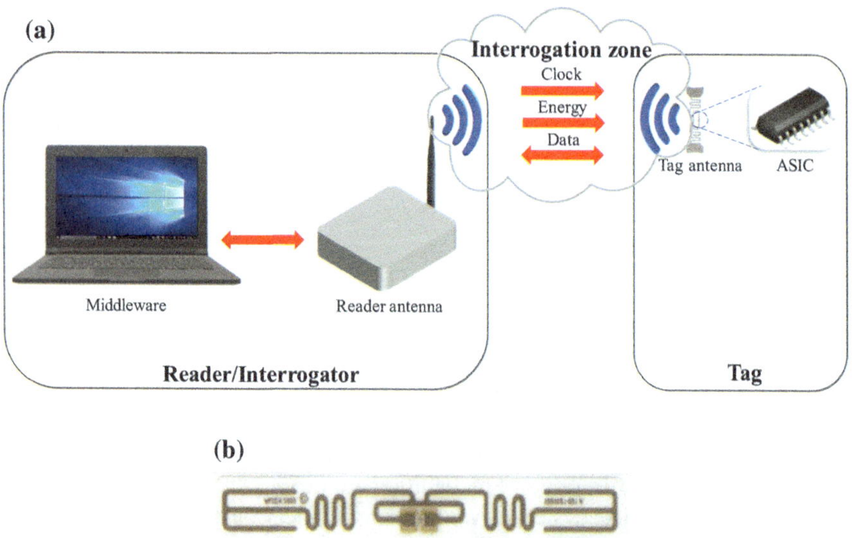

Fig. 1.1 Schematic of a typical RFID system (**a**) and photograph of a commercial UHF-RFID tag (**b**). Tag size is determined by the antenna

Table 1.1 Frequency classification of passive RFID tags

Frequency band	LF 125 and 134.2 kHz	HF 13.56 MHz	UHF 868–915 MHz	SHF 2.45 and 5.8 GHz
Physical principle	Inductive coupling	Inductive coupling	Propagation (far-field)	Propagation (far-field)
Read range	<0.5 m[a]	<1 m[a]	<15 m	<15 m[b]
Tag cost	>1€	>0.4€	>0.1€	>0.3€

[a]These are maximum read ranges but most tags operating at LF and HF exhibit read ranges of few centimeters
[b]The read range can be by far superior in active tags

silicon chips [4–21]. In chipless-RFID, the tags typically consist of a printed encoder, eventually equipped with an antenna for communication with the reader. By replacing the microchips with such encoders, tag cost can be significantly reduced, and reasonable predictions situate the cost of massively manufactured chipless-RFID tags below 1 cent [5]. Note that in chipless-RFID, tag price is dictated by the (progressively decreasing) cost of conductive inks, and by the printing fabrication processes (rotogravure, rotary or planar screen-printing, inkjet, etc.). The final cost is mainly determined by the amount of ink required by the encoder (in turn related with the data storage capability, or number of bits), and by the tag antenna, if it is present.

In terms of performance and cost, chipless-RFID tags are situated between chip-based tags and optical barcodes (or QR codes). Table 1.2 provides a comparison between these AIDC technologies that includes performance, cost, size, and other aspects, relevant in certain applications. According to the table, current challenges

Table 1.2 Comparison between chipless-RFID tags, chipped RFID tags, and optical ID codes

	Optical ID codes	Chipless RFID	Chipped RFID
Cost	Ultra low	Low[a]	Medium
Size	Very small	Large	Medium
Read range	Very small	Small/moderate	High
Data storage	Medium	Medium[a]	High
Simultaneous reading	No	No	Yes
Reprogrammable	No	Yes (with limits)	Yes
Single band operation	Yes	No[b]	Yes
Power level (reader)	–	Low	Moderate
Harsh environment	No	Yes (with limits)	No
Easy to copy	Yes	No	No

[a]Note that the considered chipless tags of the table do not include SAW based tags, where cost is not low and data storage capacity is potentially very high
[b]The exception are the chipless RFID systems (based on time-domain signature barcodes) subject of the present book, where the interrogation signal is a single-tone (harmonic) signal

in chipless-RFID research are related to the improvement in tag storage capacity, size, and read range (important limitations as compared to chipped RFID). Progress in any of these aspects, combined with the advantageous characteristics, mainly tag cost, low power needed by the reader, and robustness against harsh environments (due to the absence of electronic circuits [22]), may push chipless-RFID technology towards its progressive penetration in the market.

An important aspect highlighted in Table 1.2, related to tag costs, concerns the possibility to write and erase the tags, a feature that is not possible in optical barcodes or QR codes. By massively printing all-identical chipless tags, and programming them in a later stage, tag manufacturing costs can be reduced significantly, since a single mask is needed for that purpose. Tag programming (and erasing) is possible under certain circumstances, as will be discussed later in this book, and it is a key factor for the potential success of chipless-RFID in future. However, it is not realistic to write/erase (reprogram) the ID code of the tags as many times as in chip-based tags. Moreover, it is not always possible to manufacture all-identical tags and then write the ID code in a later stage (e.g., if the code depends on the shape of the elements forming the encoder). For these main reasons, this aspect has been market as "Yes (with limits)" in Table 1.2.

Another important issue, very sensitive in applications devoted to security and authentication, is copying and plagiarism. Optical barcodes (or QR codes) can be simply photocopied, and the copy has exactly the same functionality as the original code. Therefore, low-cost copying is possible in optical ID systems. By contrast, copying of RFID tags (both chipless and chipped) is possible, but more sophisticated (and hence high cost) systems are needed for that purpose. Particularly, chipless-RFID tags based on printed conductive inks can be reverse-engineered (unless the encoders are buried), and consequently can be reproduced. However, to this end, high-cost printers, conductive inks, and custom printing processes are required. On the other hand, although photocopied chipless-RFID tags contain, indeed, the ID code, tag reading cannot be achieved by means of a specific (dedicated) chipless-RFID reader (able to read the ink-based printed tags). Thus, counterfeiting of items or goods can be prevented, unless high-cost systems are used for copying.

Other differences between the considered wireless AIDC technologies refer to: (i) the possibility of simultaneous tag reading (only possible in chip-based RFID systems); (ii) the required power level of the reader (moderate in chipped RFID systems, as far as a minimum power level is needed to activate the chip); (iii) the bandwidth of the interrogation signal (narrow, i.e., single-band, in optical ID codes and chipped RFID systems, but wide in most chipless-RFID systems). Finally, concerning the possibility of tag operation in harsh environments, it should be mentioned that silicon ICs exhibit limited robustness against extreme ambient factors (temperature, humidity, radiation, etc.). Therefore, chipless-RFID tags can be considered to be (in general) superior in this aspect, although the properties of conductive inks may be degraded if they are subjected to extreme conditions, as well. On the other hand, in certain environments subjected to pollution, the functionality of optical barcodes and QR codes may be limited, contrary to the superior robustness of radiofrequency systems against dirtiness.

1.2 Classification of Chipless-RFID Systems

Chipless-RFID systems (or tags) can be grouped in three main categories: (i) time-domain based systems, (ii) frequency-domain based systems, and (iii) hybrid systems, where several domains are simultaneously used for tag encoding. The chipless-RFID system reported in this book can indeed be considered to belong to the first class (time-domain), but the working principle, based on time-division multiplexing and sequential bit reading by near-field coupling (see Chap. 2), is completely different. The main advantage of the chipless-RFID systems subject of this book is their data storage capacity (number of bits), only limited by tag size. The ID code is contained in the AM signal generated by the tags (a chain of functional or inoperative all-identical resonant elements or inclusions) when they are displaced over the reader, in close proximity to it (as required in a reading operation). These novel and unconventional chipless-RFID systems, based on time-domain signature barcodes (in analogy to the frequency—or spectral—signature barcodes of frequency-domain based chipless-RFID tags [5]), will be exhaustively studied in Chaps. 2 and 3. So, let us review in the remaining part of the present chapter the above-cited chipless-RFID approaches, in order to properly justify the potential of time-domain signature barcodes in terms of data capacity (the main relevant advantageous aspect of this novel chipless-RFID approach).

1.2.1 Time-Domain Chipless-RFID Systems

The former chipless-RFID systems were conceived by encoding the information in time-domain and by implementing the tags in SAW technology [23–30]. In such systems, the tags are equipped with an antenna, which captures the interrogation signal sent by the reader (typically a pulsed wave), plus an electro-acoustic transducer, responsible to transform the interrogation (electromagnetic) signal to an acoustic wave. Such wave propagates through the substrate, and experiences partial reflections at those positions where reflectors are present (note that the ID code is given by the positions occupied by the reflectors), see Fig. 1.2a. In the transducer, the reflected acoustic pulses are transformed to electromagnetic pulses and transmitted back to the reader. This working principle is generally designated as time-domain reflectometry (TDR), and the ID code is contained in the echoes generated by the reflectors. It is worth mentioning that very high data capacity SAW-based tags can be achieved (up to 256 bits, according to [29]), but at the expense of very high cost (10–20 eurocents [8]).

Alternatively, TDR-based tags can be implemented on conventional microwave substrates (or in flexible substrates) by means of delay lines with discontinuities, or other elements able to generate pulse reflections, at certain positions (Fig. 1.2b) [31–48]. As compared to the velocity of acoustic waves, typical delay lines exhibit group velocities comparable to the speed of light in vacuum. Therefore, either extremely

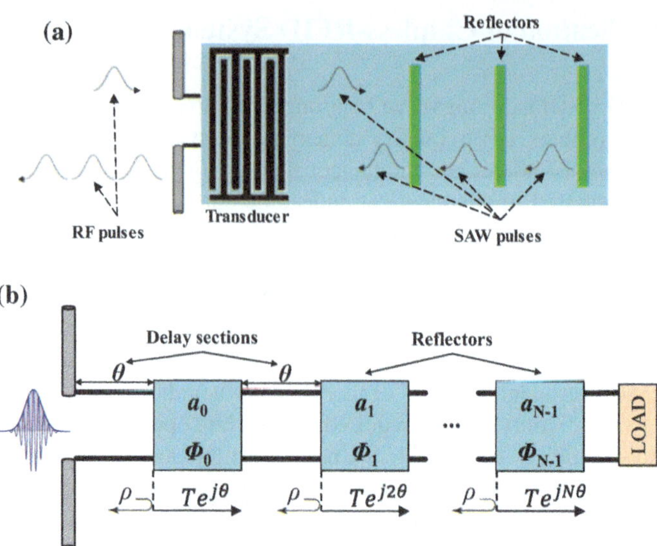

Fig. 1.2 Working principle of a TDR-based chipless RFID system. **a** Using SAW technology, **b** using delay lines

large delay lines or very narrow pulses are required in order to avoid overlapping in the reflected pulses. Since these requirements cannot be achieved in practice, bit encoding capability of TDR-based tags implemented by means of delay lines is limited. Nevertheless, let us review in the next sub-sections the most common codification techniques used in time-domain chipless-RFID systems.

1.2.1.1 ON-OFF Keying (OOK)

The most simple codification technique for TDR-based chipless-RFID systems is OOK, where the logic states '1' or '0' are determined by the presence or absence, respectively, of a reflected pulse of the interrogation signal in a given time interval.

OOK was reported by Zhang et al. in [38] by considering tags based on meandered coplanar waveguides (CPW) with capacitive discontinuities (implemented by means of surface-mount device—SMD—capacitors) inserted at predefined positions (Fig. 1.3a). The interrogation pulse (of 2 ns duration) sent by the reader propagates through the line, and it experiences a mismatching reflection provided it encounters a capacitor at a certain position in the line. Such reflection is interpreted as a logic state '1', whereas the logic state '0' is associated to the absence of such reflected pulse. An example of a measured code is depicted in Fig. 1.3b. In order to avoid pulse overlapping, the length of the delay lines must be at least 180 mm (the dielectric constant of the considered substrate is $\varepsilon_r = 3.48$). In Fig. 1.3a, four delay lines

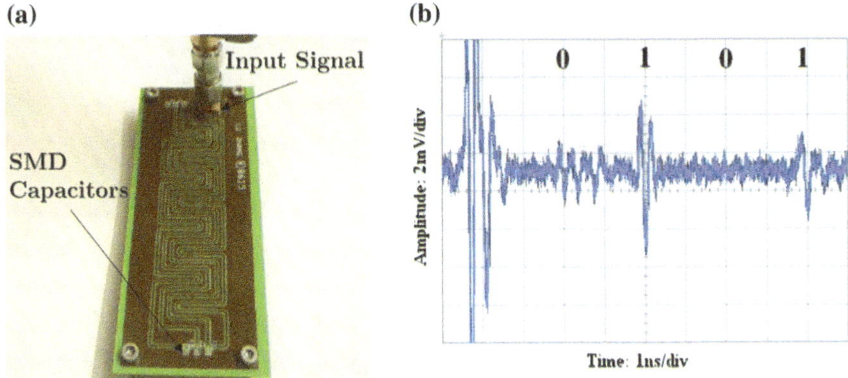

Fig. 1.3 Photograph of the chipless-RFID tag proposed by Zhang et al., with code '0101' (**a**) and measurement of such code (**b**). Reprinted with permission from [38]; copyright 2006 IEEE

occupying an area of 25 cm^2 were implemented, corresponding to a four-bit tag (one bit per delay line, in this case).

In [35, 39], the SMD capacitors were replaced with patch capacitors, as depicted in Fig. 1.4a. The patches were printed during tag fabrication, and some of them connected to the line (e.g., through inkjet printing), according to the required code, thereby making them operative. In [39], interrogation pulses of 0.4 ns duration were considered. Such pulses are compatible with 8-bit tags, provided at least 400 mm delay lines are used (the dielectric constant of the substrate considered in [39] was $\varepsilon_r = 3.48$). With these characteristics, pulse overlapping was avoided. A measured code inferred with this approach is depicted in Fig. 1.4b.

In [31, 32] a different strategy for the implementation of time-domain chipless-RFID tags based on OOK codification, with delay lines used in transmission, rather than in reflection, was reported. As it can be seen in Fig. 1.5a, the interrogation signal is divided in two branches, one being a short straight line, and the other one a meandered long line (delay line). Encoding is based on the deviation of the pulse

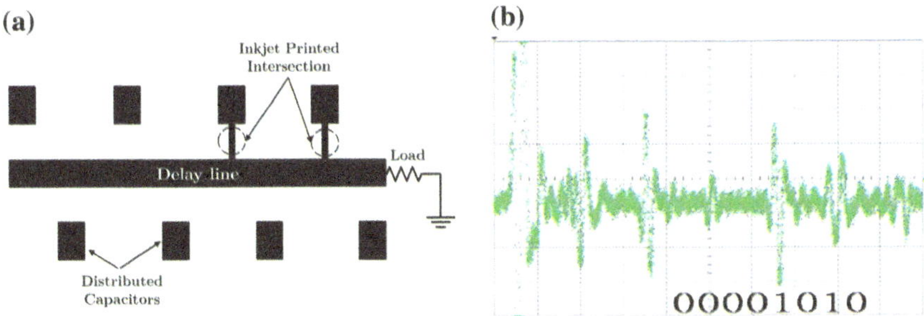

Fig. 1.4 Layout of the chipless-RFID tag reported in [39]. Corresponding to the code '00001010' (**a**) and measured code (**b**). Reprinted with permission from [39]; copyright 2008 IEEE

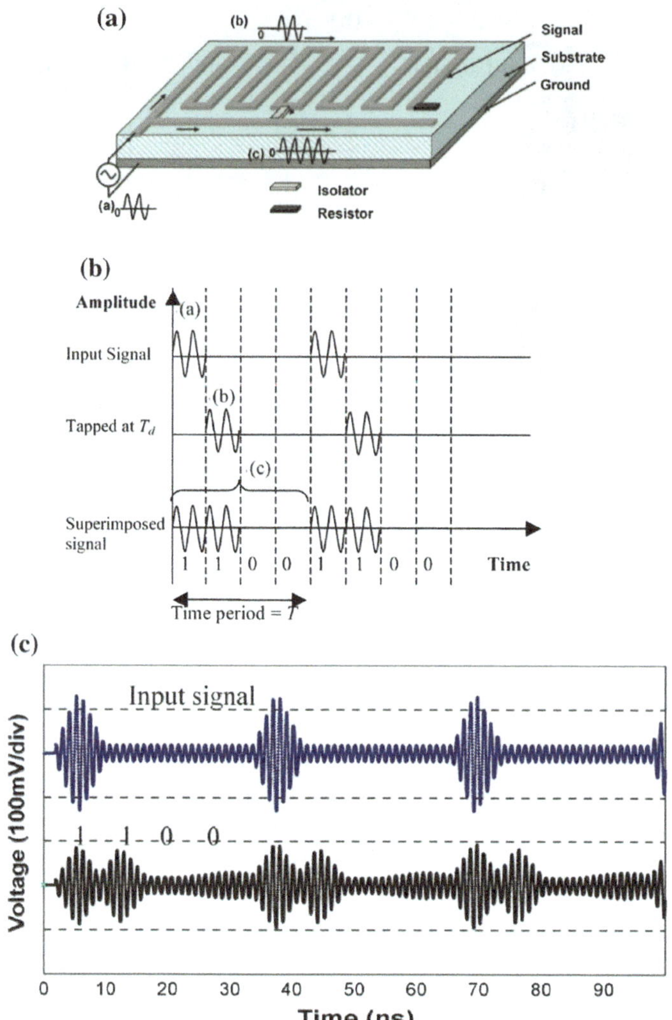

Fig. 1.5 Schematic of the time-domain transmission-based chipless-RFID tag (**a**), principle for code generation based in the superposition of delayed signals (**b**), and simulation of the input and output signals, the latter corresponding to the code '1100' (**c**). Reprinted with permission from [31]; copyright 2006 IEEE

present in the delay line to the straight line. This can be done by means of a circulator (configured as an isolator), which can be connected at different positions. Indeed, the ID code is given by the position of such circulator, which determines the delay of the deviated pulse at the output port of the straight line. Figure 1.5b, c show the principle for code generation, based on the superposition of delayed signals, and an example of measured code, respectively.

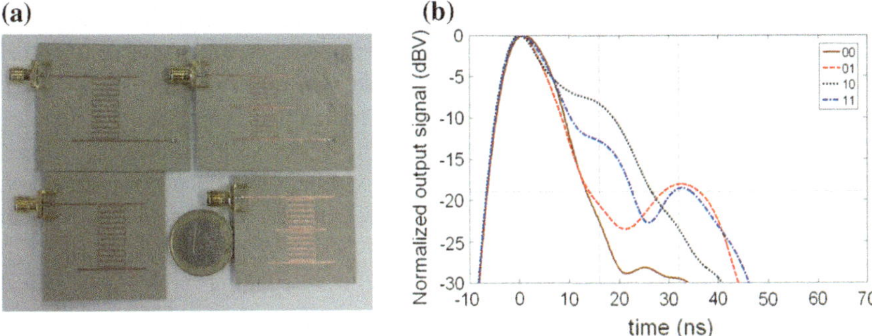

Fig. 1.6 Photograph of a set of two-bit chipless-RFID tags based on MIW delay lines (**a**) and the corresponding measured responses (**b**). Reprinted with permission from [36]; copyright 2012 IEEE

For the implementation of time-domain chipless-RFID systems based on OOK encoding, it is convenient to reduce the group velocity of the delay lines of the tags as much as possible. In this regard, an approach based on magneto-inductive wave (MIW) delay lines was reported in [36]. MIWs are slow waves that propagate along a chain of magnetically coupled resonators [49]. With such MIW delay lines, group velocities as small as $c/100$ (c being the speed of light in vacuum) can be achieved, thereby maximizing the group delay. Thus, for a given group delay (the key parameter to avoid pulse overlapping), much shorter delay lines, as compared to conventional lines, can be used for tag implementation. In [36], the considered resonators were rectangular-shaped split ring resonators (SRRs), since with this topology inter-resonator coupling is optimized. A microstrip line coupled to the first SRR of the tag chain acts as a quasi-microstrip to MIW mode transducer. Pulse reflections are generated by adding strips at certain positions of the chain, as it can be appreciated in Fig. 1.6a. Depending on the position of the reflectors, different responses, corresponding to different codes, are generated (Fig. 1.6b). The main limitation of this strategy is related to losses, which limit the number of achievable bits to a low value [36].

1.2.1.2 Pulse Position Modulation (PPM)

PPM encoding consists of varying the time position of a pulse within a certain temporal window. Specifically, PPM encodes n bits with a single pulse, which occupies one of the 2^n slots of the temporal window. The advantage of PPM encoding, as compared to OOK, is that it needs a smaller number of reflectors. However, in order to achieve the same information storage capacity, a larger temporal span is required. For instance, with a 1-ns delay line, and assuming a time resolution of 100 ps, OOK encoding provides 10 bits, whereas only 3.3 bits are achievable with PPM encoding.

Most reported works where PPM encoding is considered, use dispersive structures exhibiting a frequency dependent group velocity. Thus, a wide band signal propagating along such a dispersive structure experiences different group delay for each frequency component [42]. In practice, dispersive structures useful for PPM encoding can be implemented by means of C-section all-pass networks [43–45], consisting of a pair of coupled lines with quarter-wavelength length at the frequency where the group delay is maximum (Fig. 1.7a). The frequency position of such maximum hence depends on the length of the coupled lines (Fig. 1.7b). Consequently, for a given frequency, a pulse delay corresponding to the group delay in time domain is obtained (Fig. 1.7c).

In [44], Gupta et al. implemented a 3-bit chipless-RFID tag based on cascaded C-sections (Fig. 1.8). Such C-sections were designed in order to provide maxima in the group delay at 3, 4, and 5 GHz. In this case, each pulse may have two different positions within the temporal window, depending on the code ('0' or '1'). Additional examples of this encoding technique can be found in [16, 42, 41, 45, 46].

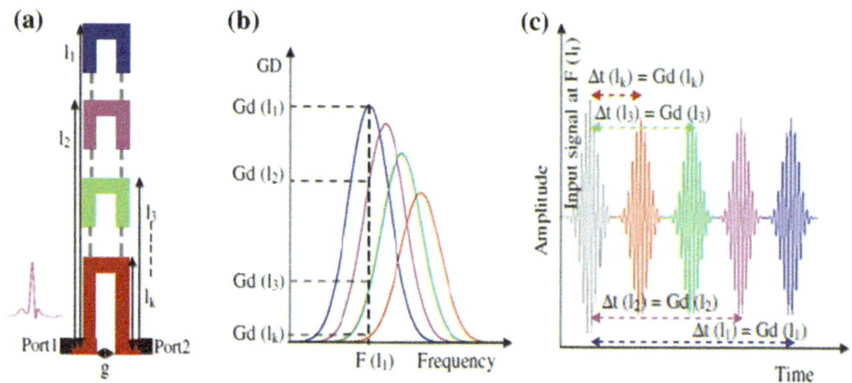

Fig. 1.7 PPM encoding principle for a chipless-RFID tag based on C-sections. **a** C-section topology, **b** group delay as a function of frequency for different C-section lengths, and **c** correspondence in time domain. Reprinted with permission from [45]; copyright 2013 Springer

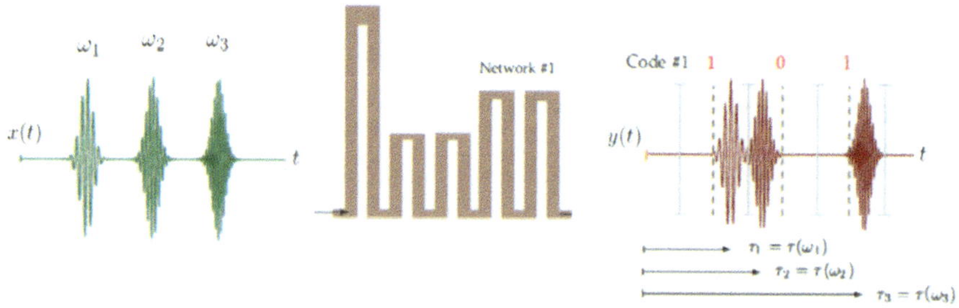

Fig. 1.8 Chipless tag proposed by Gupta et al. fed by three pulses of different frequency, ω_1, ω_2 and ω_3, and example of encoding. Reprinted with permission from [42]; copyright 2011 IEEE

1.2.1.3 Phase Modulation

Time-domain chipless-RFID systems based on phase modulation were reported in [47] as a means to increase the information density of the tags. In these systems, a delay line implemented by means of a composite right-left handed (CRLH) artificial line is used (such CRLH lines were also used in [34, 48]). The reason is that such lines can be designed in order to exhibit a small group velocity, at least as compared to the group velocity of conventional lines, thereby reducing the size of the tags for a given required delay (related to the number of bits). Different complex loads are used in order to modify the phase of the reflected pulses (Fig. 1.9). With four different elements, it is possible to achieve four different phase values (i.e., two bits for each reflector). Thus, the number of achievable codes that can be generated with these tags is 4^k, where k is the number of reflectors of the delay line.

1.2.2 *Frequency-Domain Chipless-RFID Systems*

If we exclude SAW based tags (for cost reasons), the time-domain chipless-RFID tags reported in the previous subsection exhibit very limited data capacity (the best achievement is the 8-bit tag reported in [39]). By contrast, frequency-domain based tags [5, 6, 50–68] are more competitive in terms of data storage capacity. For this main reason, most fully planar chipless-RFID tags are implemented based on frequency-domain reading, rather than time-domain. Frequency-domain based tags, also designated as spectral signature barcodes, consist of a set of resonant elements, each one tuned to a different frequency. Typically, at least in the more basic realizations of

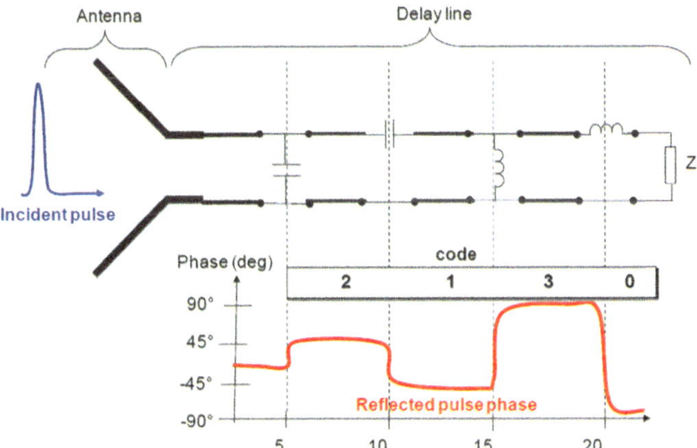

Fig. 1.9 Time-domain encoding based on phase modulation of the input pulse. Reprinted with permission from [17]; copyright 2016 Elsevier

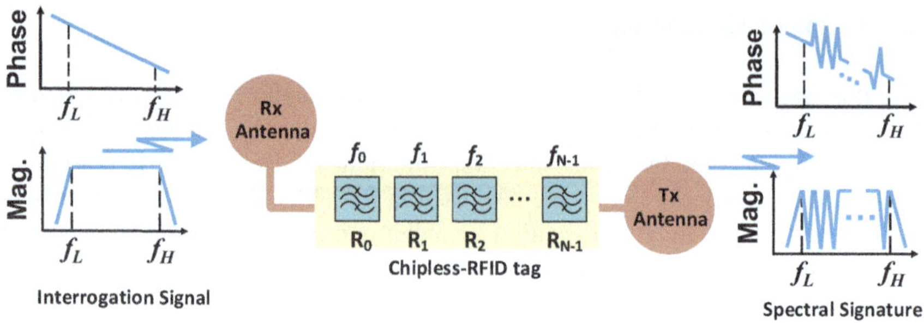

Fig. 1.10 Working principle of frequency-domain retransmission based chipless-RFID systems

these spectral signature barcodes, each resonant element provides a bit of information, and the corresponding logic state, '1' or '0', depends on whether the resonant element is functional or detuned (inoperative) at the fundamental frequency. Thus, tag reading proceeds by sending a multi-frequency interrogation signal (covering the spectral bandwidth) to the tag, and the ID code is given by the presence or absence of abrupt spectral features either in the amplitude, phase or group delay responses. According to the type of interaction between the interrogation signal and the tag, frequency-domain based tags can be classified as retransmission-based [52, 53, 60] and (ii) backscattered-based [50, 58] tags.

1.2.2.1 Retransmission-Based Tags

Figure 1.10 illustrates the working principle of retransmission-based tags, first proposed by Preradovic et al. [5, 52]. Such tags consist of a transmission line loaded with the encoding resonant elements, and equipped with two cross-polarized antennas for reception/transmission from/to the reader. It is worth mentioning the 35-bit tag reported in [5, 52], achieved by loading a meandered microstrip line with spiral resonators (Fig. 1.11). The surface and bandwidth occupied by these tags is 88 × 65 mm^2 and 4 GHz, respectively. The response of the tag with all-functional resonators (i.e., all bits set to '1') is depicted in Fig. 1.11b, where 35 dips and 35 jumps in the magnitude and phase response, respectively, are visible.

1.2.2.2 Backscattered-Based Tags

In backscattered chipless-RFID tags (see the working principle in Fig. 1.12), the resonant elements provide the spectral signature through the peaks in their radar cross section (RCS) response, and antennas are avoided, hence reducing the size of the tags. An example of tag based on this approach is shown in Fig. 1.13 [57]. Such tag consists of 20 C-shaped resonant elements with resonance frequencies comprised within the 2–4 GHz window, and tag size is 25 × 70 mm^2. As it can be appreciated in Fig. 1.13b,

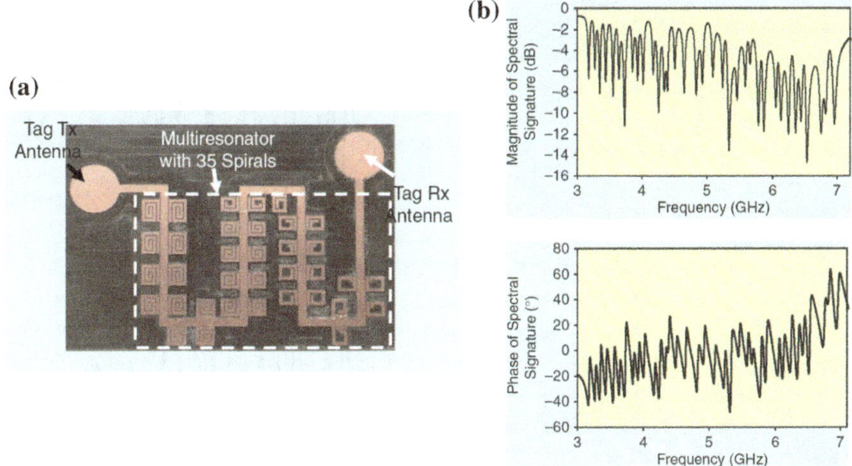

Fig. 1.11 Photograph (**a**) and measured response (**b**) of a 35-bit retransmission based chipless-RFID tag implemented by means of spiral resonators, where all bits are set to '1'. Reprinted with permission from [5]; copyright 2009 IEEE

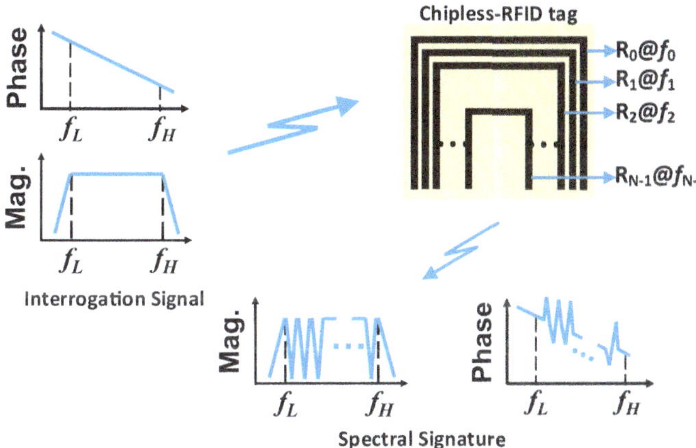

Fig. 1.12 Working principle of frequency-domain backscattered based chipless-RFID systems

peaks in both the magnitude of the RCS response and group delay response, at the resonance frequency of the functional resonators, do appear. In these tags (as well as in the tags shown in Fig. 1.11) the logic state '0' for a certain bit (or resonant element) can be generated by simply shifting the fundamental resonance frequency of the corresponding resonator to higher frequencies. In practice, such frequency shift, equivalent to make the resonant element inoperative, can be achieved by short-circuiting the resonator. Thus, it is not necessary to eliminate resonant elements of the tags in order to generate the different codes. This aspect is important as long

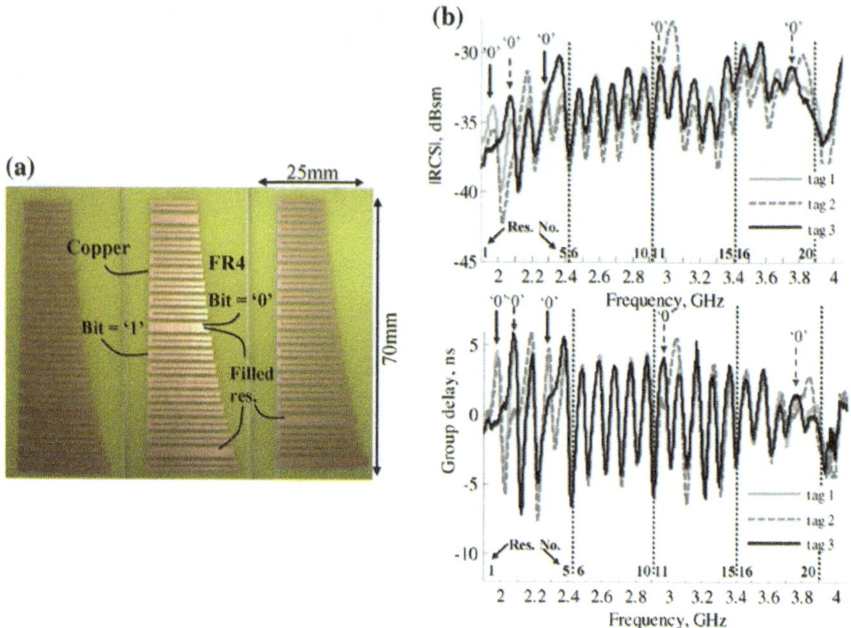

Fig. 1.13 Photograph (**a**) and measured response (**b**) of a 20-bit backscattered based chipless-RFID tag implemented by means of C-shaped resonators. Reprinted with permission from [57]; copyright 2012 IEEE

as all-identical tags (e.g., with all-functional resonators) can be fabricated, and then programmed by short-circuiting the required resonant elements (those with bit set to '0'). Nevertheless, tag programming, and erasing, will be studied in detail in Chap. 3, in reference to the time-domain signature barcodes subject of the present book.

It is possible to assign more than two logic states to a single resonator, e.g., by encoding the information in frequency [58, 59, 66]. The idea within this approach is to assign a frequency window to each resonant element, so that the logic states are given by the different (multiple) frequencies of the resonator within the interval. Obviously, it is necessary to modify the dimensions of the resonant elements in order to generate the required frequencies. Therefore, each code requires a specific mask, and tag programming with this approach is not possible. The number of states provided by each resonant element is determined by the assigned bandwidth. In order to clarify this strategy, Fig. 1.14 shows the photograph of a chipless-RFID tag based on three circular resonators [59]. Such tag operates in the ultra-wide band (UWB) frequency band, comprised between 3.1 and 10.6 GHz (corresponding to a 7.5 GHz bandwidth). By considering frequency slots of 30 MHz, the band is divided in 250 frequency slots, to share among the three ring resonators. Thus, by assigning 80 slots to each resonant element, the number of codes that can be generated is $80^3 = 512000$ codes, corresponding to roughly 19 bits (note that tag dimensions are limited to 9 cm^2, according to Fig. 1.14).

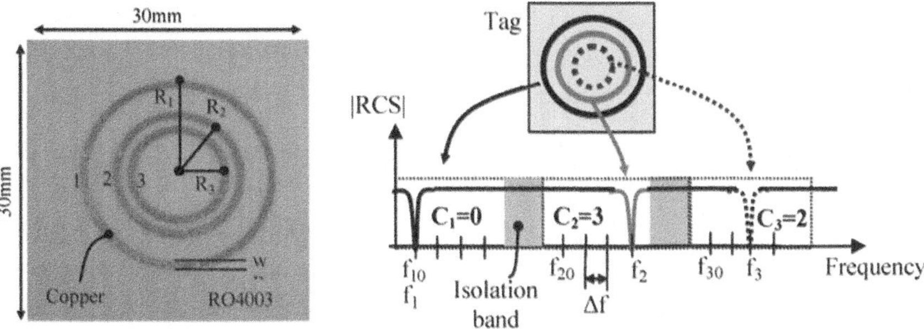

Fig. 1.14 Photograph of a chipless-RFID tag implemented by means of three circular resonators, and based on frequency position encoding. Reprinted with permission from [59]; copyright 2012 IEEE

1.2.3 Hybrid Encoding

Despite the fact that a considerable number of bits (35) has been reported in spectral signature chipless-RFID tags based on spiral resonators [5, 52] (see Fig. 1.11), the required spectral bandwidth for tag reading is very wide (this also applies to the other examples of spectral signature barcodes included in the previous sub-section or in the cited references). Such huge bandwidths represent a penalty in terms of reader costs in a real scenario, where the sweeping interrogation signal must be generated by means of a voltage-controlled oscillator (VCO). Therefore, strategies to increase the density of bits per frequency (DPF), exploiting more than one domain simultaneously, have been recently reported, giving rise to the so-called hybrid approach [69–80].

In hybrid chipless-RFID tags, the main aim is to increase the number of bits by providing more than two logic states per resonant element using simultaneously two independent parameters (or domains) for coding. By this means, the DPF and the data density per surface (DPS), two figures of merit in chipless tags, can be potentially improved. There are many proposals of hybrid tags in the literature, including tags where frequency position is combined with phase deviation [69], or with polarization diversity [70], or with notch bandwidth [79], and tags where the frequency is combined with the peak [74, 75] or notch [76, 77] magnitude, among others. Typically, hybrid tags use the frequency domain for coding. Therefore, such tags can be considered to be frequency-domain based tags, as well. Let us review some examples of hybrid chipless-RFID tags in the next subsections.

1.2.3.1 Tags Exploiting Frequency Position and Phase Deviation

In [69], Vena et al. achieved a coding capacity equivalent to 22.9 bits with only five resonant elements, by combining frequency position with phase deviation. For that purpose, the authors use C-shaped resonators, with resonance frequencies in the range 2.5–7.5 GHz, printed on *FR4* substrate (Fig. 1.15). The principle is based on

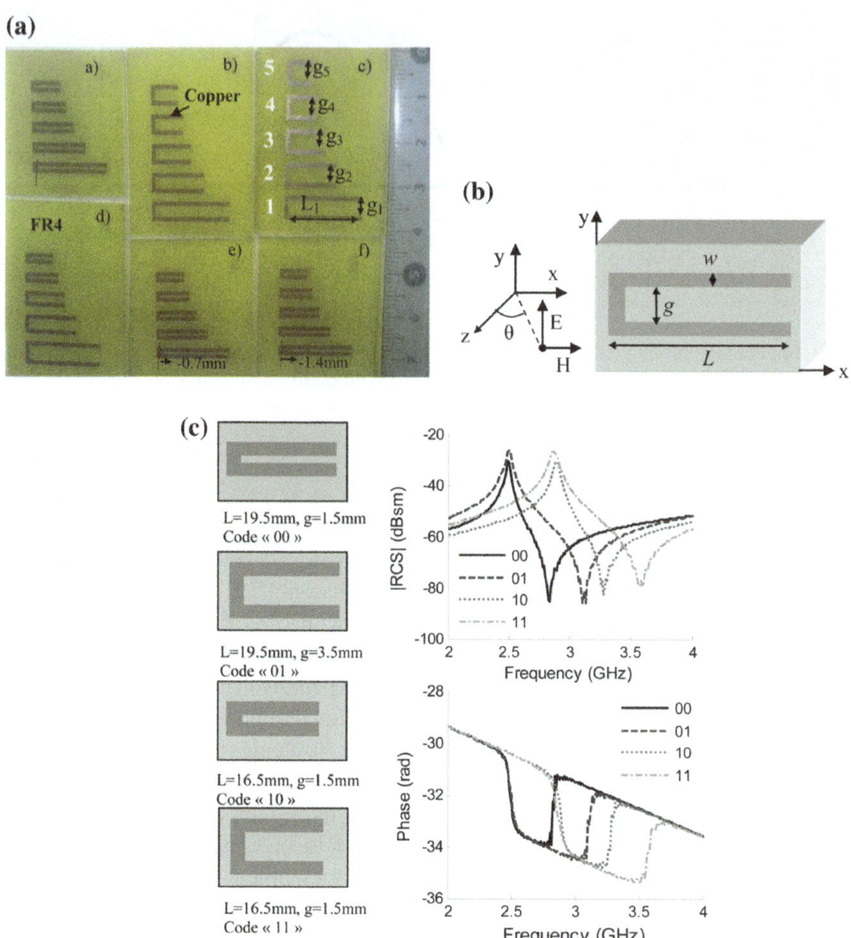

Fig. 1.15 Photograph of six chipless-RFID tags based on frequency position and phase deviation (**a**), topology of the C-shaped resonant element with relevant dimensions and polarization indicated (**b**), and responses achieved with a single C-shaped resonator (corresponding to two bits), inferred by varying the dimensions, as indicated (**c**). Reprinted with permission from [69]; copyright 2011 IEEE

the RCS responses of these resonant elements, which exhibit a peak and a dip. The peak frequency essentially depends on the length of the resonant element, whereas the dip position is given by the ratio g/L, as defined in Fig. 1.15. The figure depicts the four responses inferred from a single resonant element, by slightly modifying its dimensions. For each resonance peak, there are two dips that provide different phase response. With this example, it is shown that a single resonant element suffices to implement at least two bits. Therefore, by allowing further variations in the dimensions, it is possible to enhance the coding capacity per resonant element. In particular, in [69] the authors are able to generate 24 different states per resonant element, corresponding to 4.58 bits per resonant element.

1.2.3.2 Tags Exploiting Notch Frequency Position and Bandwidth

Hybrid encoding can be achieved by combining the frequency position of a notch and its bandwidth, as reported in [79]. The encoding technique uses three different bandwidths (BW_1, BW_2 and BW_3) and three different frequencies (f_r, $f_r + \Delta f$ and $f_r - \Delta f$), where f_r is the resonance frequency of the considered particle, and Δf is the frequency shift used for coding purposes (Fig. 1.16). In order to achieve the different bandwidths, three different encoding elements are used: a dipole, a square ring, and a patch (Fig. 1.16a). The frequency position of the notch is controlled by varying the dimensions of the resonant elements. With this strategy, it is possible to achieve a data storage capacity of 48 bits in an area of 20 cm^2, and using a frequency window of 3 GHz [79].

1.2.3.3 Tags Exploiting Frequency and Peak/Notch Magnitude

Another strategy for hybrid encoding exploits frequency and the magnitude level of either the RCS peaks (backscattered tags) or the transmission coefficient notches (retransmission-based tags). In [75], backscattered tags based on C-shaped resonators use this hybrid coding technique. The tags are very similar to those of Figs. 1.13 and 1.15, but in this case the modulation in resonator dimensions is focused on varying the peak level of the RCS. Figure 1.17 depicts three different tags, together with the measured responses (RCS magnitude level) of the three first peaks. The potential to generate multiple states per resonant element can be appreciated. Within this approach, 3 bits per resonant element are achievable, according to the authors [75].

In [77], multistate spectral signature barcodes, where encoding is carried out by varying the magnitude level of the notches associated to each resonant element, are proposed. The considered resonant elements are S-shaped split ring resonators (S-SRR) [81–83], and the notch depth is controlled by the relative orientation (by rotation) between the S-SRRs and the transmission line. By rotating the resonant element, the level of coupling between the resonator and the line varies, thereby modulating the magnitude of the notch. One limitation of this strategy is the difficulty in discerning the notch levels associated to the different states in far-field operation (by using cross-polarized antennas, as usual in this type of retransmission-based chipless-RFID tags). For this main reason, in [77], the tags are not equipped with antennas, and tag reading proceeds through near-field coupling. That is, the reader is a transmission line feed by the multi-frequency interrogation signal, and the tag should be positioned on top of the line (at short distance and conveniently aligned) in a reading operation. Figure 1.18 depicts the photograph, as well as the measured and simulated responses, of four different tags with the indicated codes. The achieved

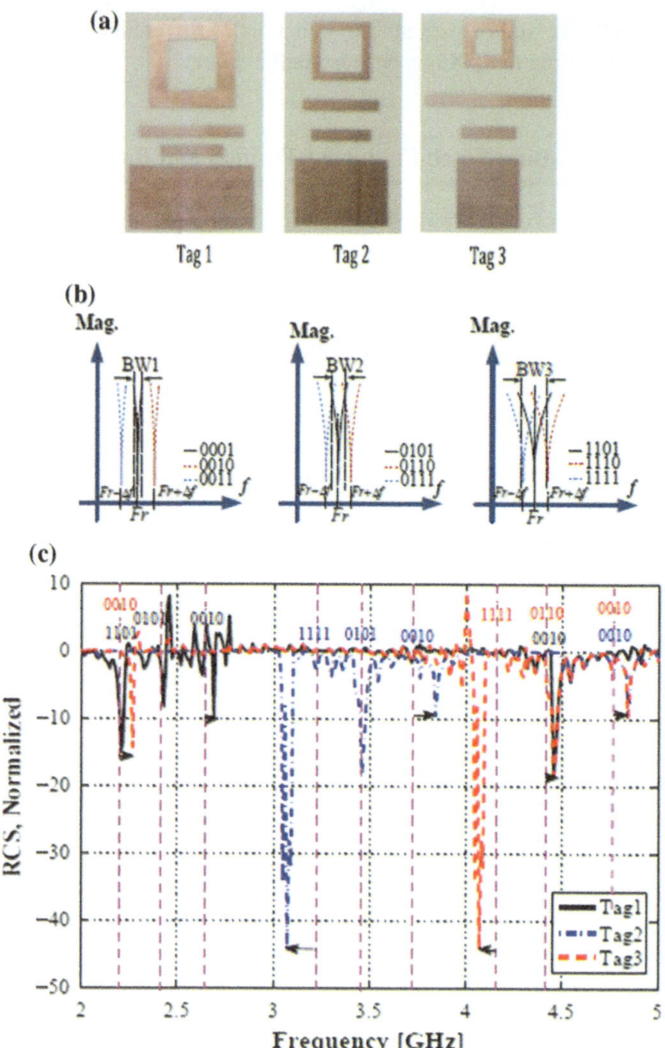

Fig. 1.16 Photograph of three chipless-RFID tags that use notch frequency position and bandwidth for coding (**a**), working principle (**b**), and normalized RCS response of the indicated tags (**c**). Reprinted with permission from [79]; copyright 2015 IEEE

DPS and DPF, 2.37 bit/cm^2 and 16 bit/GHz, respectively, are quite competitive, but at the expense of tag reading by proximity.

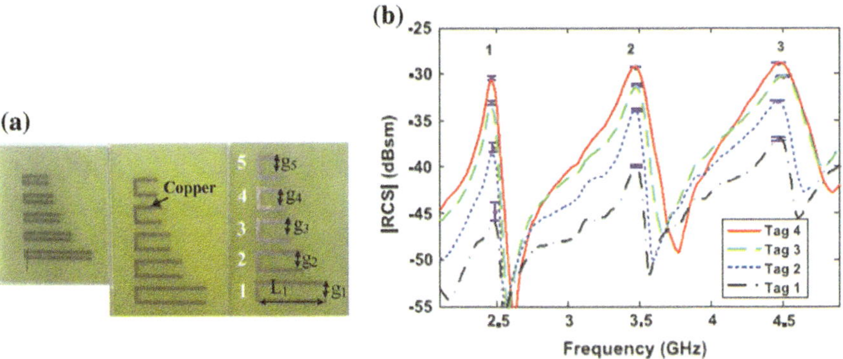

Fig. 1.17 Photograph of three chipless-RFID tags based on frequency position and RCS peak magnitude (**a**) and measured response (**b**). Reprinted with permission from [75]; copyright 2016 IEEE

1.2.3.4 Tags Exploiting Polarization Diversity

Polarization diversity is also useful to increase the information encoded in a given surface and bandwidth [70, 71]. This method is used in [71] in order to multiply by a factor of two the information capacity of the tags in the same spectral bandwidth. Such tags consist of two sets of rectangular slot resonators etched on rectangular metallic patches with different polarizations (vertical and horizontal), and, in turn, each set of resonators is divided in two patches. It is important to mention that the slots etched in each patch are conveniently placed in order to reduce the coupling between resonant elements. On the other hand, in the tag reading operation, the tag must be illuminated with a plane wave orthogonally polarized (Fig. 1.19). Therefore, the reader consists of two dual-polarized antennas, one to transmit the interrogation signal and the other one to receive the spectral signature of the tag. With this strategy, as a proof-of-concept demonstrator, the authors encoded 16 bits within an area and bandwidth of 3.06 cm^2 and 5 GHz, respectively.

1.3 Comparative Analysis and Limitations

Several approaches for the implementation of chipless-RFID systems, including time-domain encoding, frequency-domain encoding, and hybrid encoding, have been reviewed in this chapter. Table 1.3 shows a comparison of the main characteristics of several of the reported chipless-RFID systems. The most representative figure of merit is the number of bits per surface area (DPS). Nevertheless, the total number of bits is also a relevant parameter, typically limited by the occupied spectral bandwidth of the tags in frequency-domain or hybrid systems, and by the length of the delay lines and bandwidth of the pulsed interrogation signal in time-domain based tags.

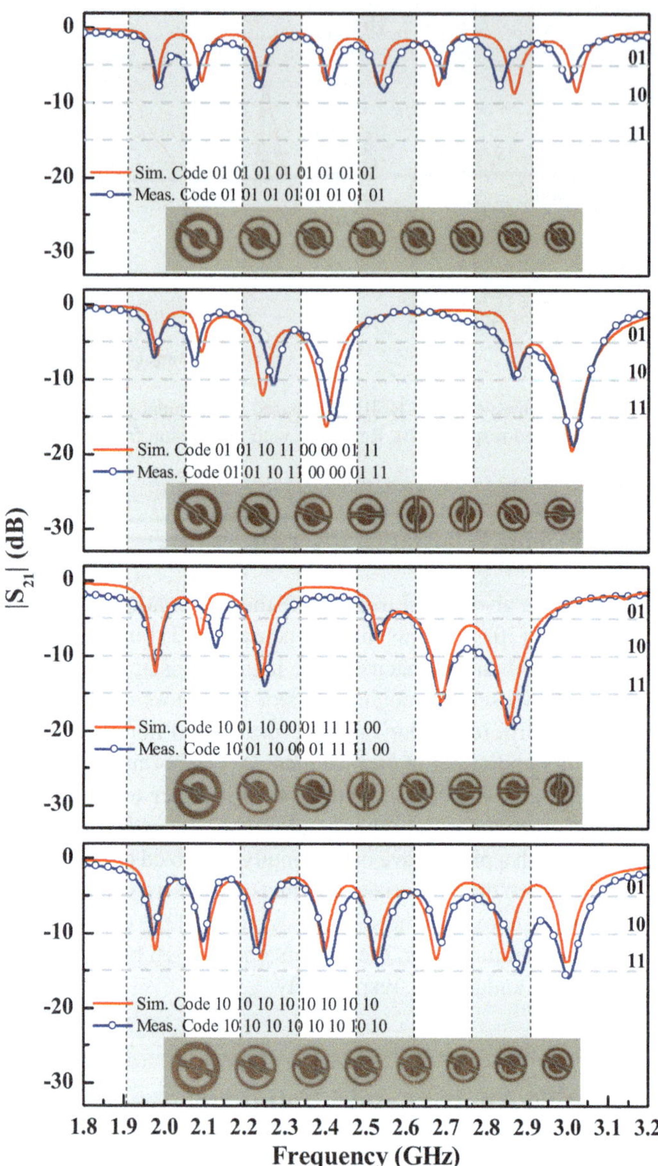

Fig. 1.18 Photograph and frequency response of the four-state eight-resonator hybrid tags exploiting frequency and notch magnitude encoding. Reprinted with permission from [77]; copyright 2017 IEEE

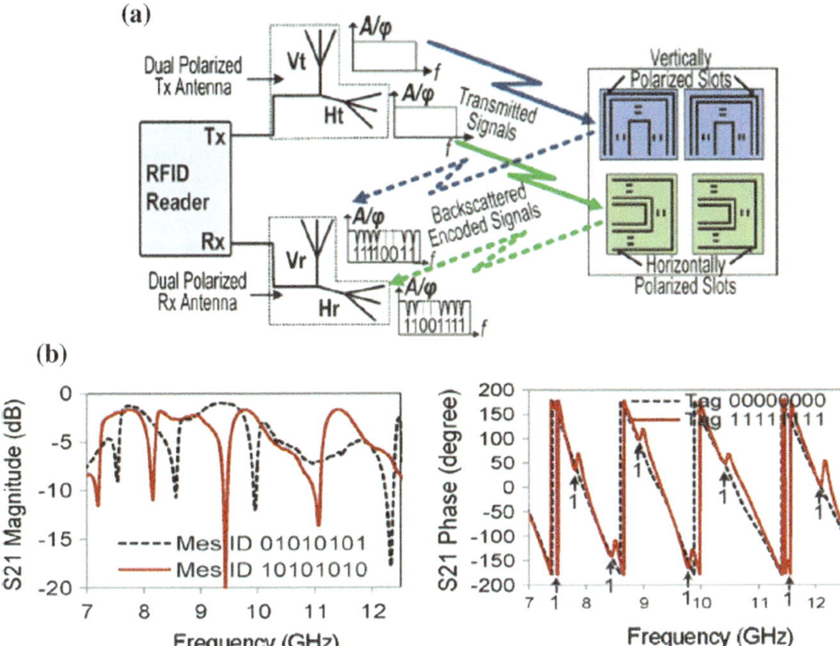

Fig. 1.19 **a** Chipless-RFID based on polarization diversity proposed by M. A. Islam et al. [71] and **b** measured frequency response of a tag with the indicated code. Reprinted with permission from [71]; copyright 2012 IEEE

In general, the spectral bandwidth of frequency-domain or hybrid systems, required to accommodate a significant number of bits, is high. Moreover, such bandwidth is proportional to the number of bits. Thus, to achieve dozens of bits (but still far from the data capacity of chipped tags), significant fractional bandwidths are needed. For instance, in [69], the fractional bandwidth occupied by the tags is as high as 100%, and only 22.9 bits are achieved, i.e., very far from the 96 bits of UHF-RFID tags, according to the electronic product code (EPC) tag data standard. In [71], 64 bits are achieved, but with a spectral bandwidth of 6.4 GHz. It is also worth mentioning the system reported in [62], where a DPS of 4.17 bit/cm^2 is achieved, but the data storage capacity of the reported tag is limited to 24 bits. The tags reported in [54] exhibit a very high DPF (as high as 25 bit/GHz), but the DPS is very small, and only 5 bits in a 50.1 cm^2 tag surface are achieved.

Note that, excluding SAW-based tags, the time-domain chipless-RFID tags reported in this review chapter exhibit a limited data capacity, as can be seen in Table 1.3. The reason is that extremely long delay lines, not of interest from a practical viewpoint, are needed in order to achieve a competitive number of bits. For this main reason, most research efforts in chipless-RFID technology have been focused on frequency-domain and hybrid encoding techniques. Nevertheless, the achieved

Table 1.3 Comparative analysis of various chipless-RFID tags

Ref.	Bandwidth (GHz)	Bits	Area (cm^2)	DPF (bit/GHz)	DPS (bit/cm^2)/DPL (bits/cm)[a]
Time-domain (Pulsed interrogation signal)					
[31]	–	4	59.4	–	0.07/0.02
[36]	0.05	2	–	40	– /0.10
[38]	–	4	–	–	– /0.06
[39]	–	8	–	–	– /0.20
[40]	0.8	5	26	6.25	0.19/0.19
[41]	–	2	70.0	–	0.03/–
[46]	–	2	8.75	–	0.23/–
Frequency-domain					
[50]	1	5	6.48	11.1	0.77
[52]	7.5	35	57.2	8.97	0.61
[54]	0.2	5	50.1	25.0	0.10
[57]	2	20	17.5	10.0	1.14
[58]	3.5	9	3.00	2.57	3.00
[59]	7.5	19	9.00	2.53	2.11
[61]	7	28.5	8.00	4.07	3.56
[62]	7.5	24	5.76	3.20	4.17
[67]	1.2	20	17.5	16.7	1.14
Hybrid					
[69]	5	22.9	8.00	4.58	2.86
[71]	6.4	64	10.9	10.0	5.88
[77]	1	16	6.75	16.0	2.37
[75]	3	9	7.20	3.00	1.25

[a]In time-domain tags, the density of bits per unit length (DPL) is sometimes given, especially in tags based on delay lines

tag performance (mainly data capacity and size) is still far from the one of chipped-based tags. By contrast, in the time-domain approach reported in this book (Chaps. 2 and 3), where tags are read by time-division multiplexing through a harmonic (i.e., single-tone) interrogation signal, the achievable number of bits may compete against the data storage capacity of chipped tags, yet keeping relative small tag sizes. Tag reading by proximity (near-field) is a due in this unconventional chipless-RFID system, but this is not necessarily an issue in many applications, especially those related to security and authentication (the most relevant, but not exclusive, application of the reported approach).

References

1. Finkenzeller K, Handbook RFID (2004) Radio-frequency identification fundamentals and applications, 2nd edn. John Wiley, New York, USA
2. Hunt VD, Puglia A, Puglia M (2007) RFID: a guide to radio frequency identification. John Wiley, New York, USA
3. Finkenzeller K (2010) RFID handbook: fundamentals and applications in contactless smart cards, radio frequency identification and near-field communications, 3rd edn. John Wiley, Hoboken, NJ, USA
4. Karmakar NC, Kalansuriya P, Azim RE, Koswatta R (2016) Chipless radio frequency identification reader signal processing. John Wiley, Hoboken, NJ, USA
5. Preradovic S, Karmakar NC (2010) Chipless RFID: bar code of the future. IEEE Microw Mag 11:87–97
6. Preradovic S, Karmakar NC (2011) Multiresonator-based chipless RFID: barcode of the future. Springer-Verlag, New York, USA
7. Karmakar NC, Koswatta R, Kalansuriya P, E-Azim R (2013) Chipless RFID reader architecture. Artech House
8. Perret E (2014) Radio frequency identification and sensors: from RFID to chipless RFID. John Wiley, New York, USA
9. Rezaiesarlak R, Manteghi M (2015) Chipless RFID: design procedure and detection techniques. Springer
10. Karmakar NC, Zomorrodi M, Divarathne C (2016) Advanced chipless RFID. John Wiley, Hoboken, NJ, USA
11. Tedjini S, Karmakar N, Perret E, Vena A, Koswatta R, E-Azim R (2013) Hold the chips: chipless technology, an alternative technique for RFID. IEEE Microw Mag 14(5):56–65
12. Karmakar NC (2016) Tag, You're it radar cross section of chipless RFID tags. IEEE Microw Mag 17(7):64–74
13. Dey S, Saha JK, Karmakar NC (2015) Smart sensing: chipless RFID solutions for the internet of everything. IEEE Microw Mag 16(10):26–39
14. Shao B, Chen Q, Amin Y, Liu R, Zheng L-R (2013) Chipless RFID tags fabricated by fully printing of metallic inks. Ann Telecommun-Ann Télécommunications 68(7–8):401–413
15. Vena A, Perret E, Tedjini S (2013) Design rules for chipless RFID tags based on multiple scatterers. Ann Telecommun-Ann Télécommunications 68(7–8):361–374
16. Forouzandeh M, Karmakar NC (2015) Chipless RFID tags and sensors: a review on time-domain techniques. Wirel Power Transf 2(2):62–77
17. Vena A, Perret E, Tedjini S (2016) Chipless RFID based on RF encoding particle: realization, coding and reading system. ISTE Press – Elsevier
18. Karmakar NC, Amin EM, Saha JK (2016) Chipless RFID sensors. Wiley, Hoboken, NJ
19. Rance O, Perret E, Siragusa R, Lemaitre-Auger P (2017) RCS synthesis for chipless RFID: theory and design. Elsevier
20. Herrojo C, Moras M, Paredes F, Núñez A, Mata-Contreras J, Ramon E, Martín F (2019) Time-domain signature chipless-RFID tags: near-field chipless-RFID systems with high data capacity. IEEE Microw Mag 20(12):87–101
21. Herrojo C, Paredes F, Mata-Contreras J, Martín F (2019) Chipless-RFID: a review and recent developments. Sensors 19(15):3385
22. Moscato S, et al. (2014) Chipless RFID for space applications. In: 2014 IEEE international conference on wireless for space and extreme environments (WiSEE), Noordwijk, Netherlands, Oct 2014
23. Nysen PA, Skeie H, Armstrong D (1988) System for interrogating a passive transponder carrying phase-encoded information. Google Patents
24. Hartmann CS (2002) A global SAW ID tag with large data capacity. In: Proceedings of IEEE ultrasonics symposium, vol 1, pp 65–69
25. Saldanha N, Malocha DC (2007) Design Parameters for SAW multi-tone frequency coded reflectors. In: 2007 IEEE ultrasonics symposium, pp 2087–2090

26. Harma S, Plessky VP, Hartmann CS, Steichen W (2008) Z-path SAW RFID tag. IEEE Trans Ultrason Ferroelectr Freq Control 55:208–213
27. Han T, Wang W, Wu H, Shui Y (2008) Reflection and scattering characteristics of reflectors in SAW tags. IEEE Trans Ultrason Ferroelectr Freq Control 55(6):1387–1390
28. Harma S, Plessky VP, Li X, Hartogh P (2009) Feasibility of ultra-wideband SAW RFID tags meeting FCC rules. IEEE Trans Ultrason Ferroelectr Freq Control 56:812–820
29. Hartmann C, Hartmann P, Brown P, Bellamy J, Claiborne L, Bonner W (2004) Anti-collision methods for Global SAW RFID Tag systems. IEEE Ultrason Symp 2:805–808
30. Tao H, Weibiao W, Haodong W, Yongan S (2008) Reflection and scattering characteristics of reflectors in SAW tags. IEEE Trans Ultrason, Ferroelectr Freq Control 55:1387–1390
31. Chamarti A, Varahramyan K (2006) Transmission delay line based ID generation circuit for RFID applications. IEEE Microw Wireless Compon Lett 16:588–590
32. Vemagiri J, Chamarti A, Agarwal M, Varahramyan K (2007) Transmission line delay-based radio frequency identification (RFID) tag. Microw Opt Technol Lett 49(8):1900–1904
33. Schüßler M, Damm C, Jakoby R (2007) Periodically LC loaded lines for RFID backscatter applications. Metamaterials 2007, Rome, Italy, pp 103–106, Oct 2007
34. Schüßler M, Damm C, Maasch M, Jakoby R Performance evaluation of left-handed delay lines for RFID backscatter applications. In: IEEE MTT-S international microwave symposium 2008, Atlanta, GA, USA, pp 177–180
35. Shao B, Chen Q, Amin Y, Mendoza DS, Liu R, Zheng L-R (2010) An ultra-low-cost RFID tag with 1.67 Gbps data rate by ink-jet printing on paper substrate. In: IEEE asian solid state-circuits conference, Beijing, China, pp 1–4
36. Herraiz-Martínez FJ, Paredes F, Zamora G, Martín F, Bonache J (2012) Printed magnetoinductive-wave (MIW) delay lines for chipless RFID applications. IEEE Trans Ant Propag 60:5075–5082
37. Tedjini S, Perret E, Vena A, Kaddout D (2012) Mastering the electromagnetic signature of chipless RFID tags. In: Chipless and conventional radiofrequency identification, ed. IGI Global
38. Zhang L, Rodriguez S, Tenhunen H, Zheng L-R An innovative fully printable RFID technology based on high speed time-domain reflections. In: Conference on high density microsystem design and packaging and component failure analysis, 2006. HDP'06, Shanghai, China, June 2006, pp 166–170
39. Zheng L, Rodriguez S, Zhang L, Shao B, Zheng L-R (2008) Design and implementation of a fully reconfigurable chipless RFID tag using Inkjet printing technology. In: 2008 IEEE international symposium on circuits and systems, Seattle, USA, May 2008, pp 1524–1527
40. Mandel C, Schussler M, Maasch M, Jakoby R (2009) A novel passive phase modulator based on LH delay lines for chipless microwave RFID applications. In: 2009 IEEE MTT-S international microwave workshop on wireless sensing, local positioning, and RFID, Cavtat, Croatia, Sep 2009, pp 1–4
41. Nair R, Perret E, Tedjini S (2012) Temporal multi-frequency encoding technique for chipless RFID applications. In: IEEE MTT-S international microwave symposium, Montreal, Canada, June 2012, pp 1–3
42. Gupta S, Nikfal B, Caloz C (2010) RFID system based on pulse-position modulation using group delay engineered microwave C-sections. In: 2010 Asia-Pacific microwave conference, Yokohama, Japan, Dec 2010, pp 203–206
43. Cristal EG (1966) Analysis and exact synthesis of cascaded commensurate transmission-line C-Section all-pass networks. IEEE Trans Microw Theory Tech 14(6):285–291
44. Gupta S, Nikfal B, Caloz C (2011) Chipless RFID system based on group delay engineered dispersive delay structures. IEEE Antennas Wirel Propag Lett 10(2):1366–1368
45. Nair RS, Perret E, Tedjini S (2013) Group delay modulation for pulse position coding based on periodically coupled C-sections. Ann Telecommun 68(7–8):447–457
46. Nair R, Perret E, Tedjini S (2011) Chipless RFID based on group delay encoding. In: 2011 IEEE international conference on RFID-technologies and applications, Sitges, Spain, vol 1, Sep 2011, pp 214–218

47. Mandel C, Schussler M, Maasch M, Jakoby R (2009) A novel passive phase modulator based on LH delay lines for chipless microwave RFID applications. In: 2009 IEEE MTT-S international microwave workshop on wireless sensing, local positioning, and RFID, Sep 2009, pp 1–4

48. Schussler M, Mandel C, Maasch M, Giere A, Jakoby R (2009) Phase modulation scheme for chipless RFID- and wireless sensor tags. In: 2009 Asia Pacific microwave conference, Dec 2009, Singapore, pp. 229–232

49. Shamonina E, Kalinin VA, Ringhofer KH, Solymar L (2002) Magneto-inductive waveguide. Electron Lett 38:371–373

50. Jalaly I, Robertson ID (2005) RF barcodes using multiple frequency bands. In: IEEE MTT-S international microwave symposium, Long Beach, USA, June 2005, pp 139–142

51. Preradovic S, Balbin I, Karmakar NC, Swiegers G (2008) A novel chipless RFID system based on planar multiresonators for barcode replacement. In: 2008 IEEE International Conference on RFID, Las Vegas, USA, Apr 2008, pp 289–296

52. Preradovic S, Balbin I, Karmakar NC, Swiegers GF (2009) Multiresonator-based chipless RFID system for low-cost item tracking. IEEE Trans Microw Theory Techn 57: 1411–1419

53. Preradovic S, Karmakar NC (2010) Design of chipless RFID tag for operation on flexible laminates. IEEE Anten Wireless Propag Lett 9:207–210

54. McVay J, Hoorfar A, Engheta N (2006) Space-filling curve RFID tags. IEEE Radio Wireless Symp, San Diego, CA, USA, pp 199–202

55. Jalaly I, Robertson D (2005) Capacitively-tuned split microstrip resonators for RFID barcodes. In: 2005 European microwave conference, Paris, France, vol 2, Oct 2005, pp 4–7

56. Jang H-S, Lim W-G, Oh K-S, Moon S-M, Yu J-W (2010) Design of low-cost chipless system using printable chipless tag with electromagnetic code. IEEE Microw Wireless Compon Lett 20:640–642

57. Vena A, Perret E, Tedjini S (2012) A fully printable chipless RFID tag with detuning correction technique. IEEE Microw Wireless Compon Lett 22(4):209–211

58. Vena A, Perret E, Tedjini S (2012) Design of compact and auto-compensated single-layer chipless RFID tag. IEEE Trans Microw Theory Techn 60(9):2913–2924

59. Vena A, Perret E, Tedjini S (2012) High-capacity chipless RFID tag insensitive to the polarization. IEEE Trans Ant Propag 60(10):4509–4515

60. Girbau D, Lorenzo J, Lazaro A, Ferrater C, Villarino R (2012) Frequency-coded chipless RFID tag based on dual-band resonators. IEEE Ant Wireless Propag Lett 11:126–128

61. Khan MM, Tahir FA, Farooqui MF, Shamim A, Cheema HM (2016) 3.56-bits/cm^2 compact inkjet printed and application specific chipless RFID tag. IEEE Ant Wireless Propag Lett 15:1109–1112

62. Rezaiesarlak R, Manteghi M (2014) Complex-natural-resonance-based design of chipless RFID tag for high-density data. IEEE Trans Ant Propag 62:898–904

63. Bhuiyan MS, Karmakar N (2014) A spectrally efficient chipless RFID tag based on split-wheel resonator. Int. Antenna Technol. Workshop on Small Antennas, Novel EM Struct., Mater., Appl., pp 1–4

64. Nijas CM et al (2014) Low-cost multiple-bit encoded chipless RFID tag using stepped impedance resonator. IEEE Trans Ant Propag 62(9):4762–4770

65. Machac J, Polivka M (2014) Influence of mutual coupling on performance of small scatterers for chipless RFID tags. In: 24th Int Radioelektron Conf, pp. 1–4

66. Khaliel M, El-Awamry A, Fawky A, El-Hadidy M, Kaiser T (2015) A novel co/cross-polarizing chipless RFID tags for high coding capacity and robust detection. In: 2015 IEEE international symposium on antennas and propagation & USNC/URSI national radio science meeting, Jul 2015, pp 159–160

67. Svanda M, Machac J, Polivka M, Havlicek J (2016) A comparison of two ways to reducing the mutual coupling of chipless RFID tag scatterers. In: Proceedings of 21st International Conference on Microwave, Radar and Wireless Communications (MIKON), May 2016, pp 1–4

68. Herrojo C, Naqui J, Paredes F, Martín F (2016) Spectral signature barcodes based on S-shaped Split ring resonators (S-SRR). EPJ Appl Metamaterials 3:1–6

69. Vena A, Perret E, Tedjini S (2011) Chipless RFID tag using hybrid coding technique. IEEE Trans Microw Theory Techn 59:3356–3364
70. Vena A, Perret E, Tedjini S (2012) A compact chipless RFID tag using polarization diversity for encoding and sensing. In: 2012 IEEE International Conference on RFID, pp 191–197
71. Islam MA, Karmakar NC (2012) A novel compact printable dual-polarized chipless RFID system. IEEE Trans Microw Theory Techn 60:2142–2151
72. Balbin I, Karmakar NC (2009) Phase-encoded chipless RFID transponder for large scale low cost applications. IEEE Microw Wireless Compon Lett 19:509–511
73. Genovesi S, Costa F, Monorchio A, Manara G (2015) Chipless RFID tag exploiting multifrequency delta-phase quantization encoding. IEEE Ant Wireless Propag Lett 15:738–741
74. Rance O, Siragusa R, Lemaitre-Auger P, Perret E (2015) RCS magnitude coding for chipless RFID based on depolarizing tag. In: IEEE MTT-S International Microwave Symposium Digest, Phoenix, AZ, USA, pp 1–4
75. Rance O, Siragusa R, Lemaître-Auger P, Perret E (2016) Toward RCS magnitude level coding for chipless RFID. IEEE Trans Microw Theory Techn 64:2315–2325
76. Herrojo C, Naqui J, Paredes F, Martín F (2016) Spectral signature barcodes implemented by multi-state multi-resonator circuits for chipless RFID tags. In: IEEE MTT-S International Microwave Symposium (IMS'16), San Francisco, May 2016
77. Herrojo C, Paredes F, Mata-Contreras J, Zuffanelli S, Martín F (2017) Multi-state multi-resonator spectral signature barcodes implemented by means of S-shaped Split Ring Resonators (S-SRR). IEEE Trans Microw Theory Techn 65(7):2341–2352
78. Feng C, Zhang W, Li L, Han L, Chen X, Ma R (2015) Angle-based chipless RFID tag with high capacity and insensitivity to polarization. IEEE Trans Ant Propag 63(4):1789–1797
79. El-Awamry A, Khaliel M, Fawky A, El-Hadidy M, Kaiser T (2015) Novel notch modulation algorithm for enhancing the chipless RFID tags coding capacity. In: IEEE International Conference on RFID, San Diego, CA, USA, pp 25–31
80. Vena A, Babar AA, Sydanheimo L, Tentzeris MM, Ukkonen L (2013) A novel near-transparent ASK-reconfigurable inkjet-printed chipless RFID tag. IEEE Ant Wireless Propag Lett 12:753–756
81. Chen H, Ran L, Huangfu J, Zhang X, Chen K, Grzegorczyk TM, Kong JA (2004) Left-handed materials composed of only S-shaped resonators. Phys Rev E 70(5):057605
82. Chen H, Ran L, Huangfu J, Zhang X, Chen K, Grzegorczyk TM, Kong JA (2005) Negative refraction of a combined double S-shaped metamaterial. Appl Phys Lett 86(15):151909
83. Naqui J, Coromina J, Karami-Horestani A, Fumeaux C, Martín F (2015) Angular displacement and velocity sensors based on coplanar waveguides (CPWs) loaded with S-shaped split ring resonator (S-SRR). Sensors 15:9628–9650

Chapter 2
Time-Domain Signature Near-Field Chipless-RFID Systems

A different working principle for the implementation of chipless-RFID systems, based on the time domain and near-field coupling between the tag and the reader, is reported and discussed in this chapter. As it will be shown, proximity and proper alignment between the tag and the reader is required for tag reading in such chipless-RFID systems. However, their data storage capacity is only limited by tag size, and it is possible to implement reasonably sized tags with very competitive number of bits, comparable to the number of bits of commercial chipped-RFID systems. Although this chapter is devoted to provide a detailed description of the general working principle of these chipless-RFID systems, to discuss the different strategies to achieve a high data density (a figure of merit), and to report illustrative examples based on different configurations, it is worth mentioning that there are various scenarios where such systems may find practical application, including secure paper, motion control, and chipless-RFID sensing, among others (such applications, however, will be discussed in Chapter 3).

2.1 Working Principle and Advantages/Limitations as Compared to Other Chipless-RFID Systems

In the frequency-domain and hybrid chipless-RFID systems discussed in Chapter 1, resonant elements tuned to different frequencies are used for coding purposes (see [1] and references therein, or the references of Chapter 1). In such systems, tag reading requires a wideband interrogation signal able to cover the whole spectral bandwidth occupied by the resonant elements of the tag. Despite the fact that the DPF in such spectral signature barcodes, as they are usually designated, can be enhanced by using various domains simultaneously (hybrid tags provide more than one bit of information per resonant element [2–4]), the reported data capacity to date is far from the number of bits of chipped-RFID systems (96 bits according to the EPC class 1 generation 2 protocol in UHF-RFID [5]), see Chapter 1 for further details.

© Springer Nature Switzerland AG 2020

F. Martín et al., *Time-Domain Signature Barcodes for Chipless-RFID and Sensing Applications*, Lecture Notes in Electrical Engineering 647,
https://doi.org/10.1007/978-3-030-39726-5_2

This limited data capacity of frequency-domain and hybrid chipless-RFID systems is related to the required bandwidth of the interrogation signal, a multi-frequency signal that must be swept over the spectral bandwidth of the tags. Such wideband interrogation signals, with fractional bandwidths roughly proportional to the number of bits of the tags, are not compatible with low-cost readers, as far as their generation requires complex front-ends based on wideband VCOs, thereby limiting the data capacity of the tags.

Tag reading in frequency-domain and hybrid chipless-RFID systems proceeds, indeed, by frequency-division multiplexing, where the tag ID code is frequency distributed, i.e., different bits are assigned to different frequencies. Obviously, the ID codes are also distributed in space, as far as the resonant elements of the tags are printed at different positions and occupy a finite area.[1] The chipless-RFID approach presented in this chapter is conceptually similar to frequency-domain encoding, but it uses time-division multiplexing for tag reading [6]. Thus, the tag consists of a set of all-identical inclusions (typically resonant elements) distributed in space, where each inclusion provides a bit of information and it is printed, or etched, in a predefined position in the tag. The logic state '1' or '0' associated to each bit is determined by the presence or absence, respectively, of a functional (operative) inclusion in the position corresponding to that bit. It should be clarified that, within the context of this book, a functional inclusion is the one that exhibits its properties at the intended frequency. For resonant inclusions, functionality requires that the resonant element resonates at the design frequency of the tag (dictated by the reader, as it will be later shown), provided it is properly excited. Therefore, the '0' state for a certain bit (associated to a predefined tag position) can be achieved, e.g., by the absence of inclusion in that position [6, 7]. For resonant inclusions, by detuning a resonant element (e.g., by cutting or by short-circuiting it, depending on the specific resonator topology), it is made inoperative at the design frequency, thereby providing the same logic state ('0') [1, 8–12]. Within this approach, the ID code is space-distributed, and tag reading requires a system able to detect the functional inclusions and their position in the tag, as this provides unequivocally the ID code.

From now on, let us consider, without loss of generality, that the inclusions are resonant elements, as most time-domain signature barcodes are based on resonators.[2] Since the resonant elements of the tag are identical, i.e., all tuned to the same frequency, the spectral bandwidth of the tags is virtually null. Hence, a single-tone (harmonic) signal can be considered as interrogation signal in order to detect the functionality (resonance) of the resonant elements in a reading operation [1, 6–12]. For that purpose, it is convenient to arrange the resonant elements of the tag forming a periodic chain configuration (with the resonators printed, or etched, at predefined and equidistant positions—Fig. 2.1), and to read the tags sequentially, bit by bit at

---

[1]Note that, by contrast, in chip-based RFID systems the ID codes are "dimensionless", and the bandwidth of the interrogation signal is very narrow.

[2]Some chipless-RFID tags based on linear strips (to be discussed in Sect. 2.3.3) do not behave as resonant chains, but the working principle for tag reading is identical, i.e., the AM modulation of the harmonic interrogation signal.

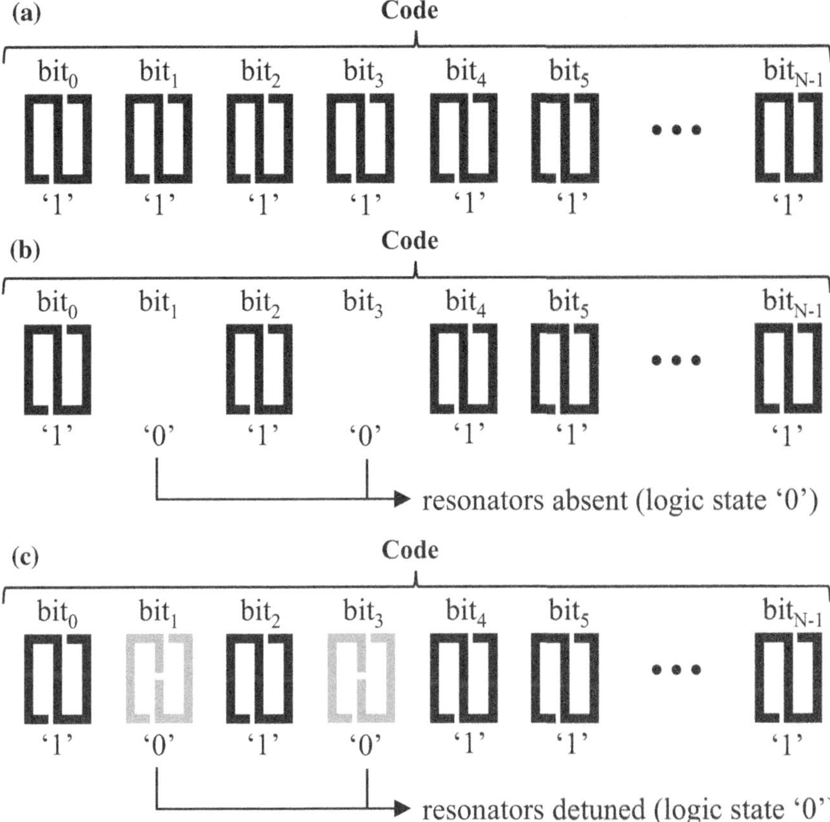

Fig. 2.1 Linear chain of resonant elements corresponding to a time-domain signature barcode with all resonant elements present and functional (**a**), with some resonators absent (**b**), and with all resonant elements present but some of them detuned (**c**). The corresponding bits are indicated. Note that the same effect ('0') is achieved by the absence of resonant element or by resonator detuning. The considered resonant elements are S-SRRs

different time, through near-field. Clearly, this scheme is based on time-division multiplexing, and the ID code is contained in the response of the tag to the interrogation signal, obtained at different (predefined) times. For this reason, in analogy to spectral (or frequency-domain) signature barcodes, the tags of this unconventional chipless-RFID approach, based on near-field and sequential bit reading, can be designated as time-domain signature barcodes (the ID code is read in time domain) [1].

As compared to frequency-domain or hybrid systems, and taking into account that a simple harmonic signal suffices to read these time-domain signature barcodes (rather than a multi-frequency interrogation signal), the data capacity of these near-field time-division multiplexing chipless-RFID systems can be substantially increased. Indeed, the limitation may be dictated by the space occupied by the tags, but not by the complexity of the reading system, which only involves the generation

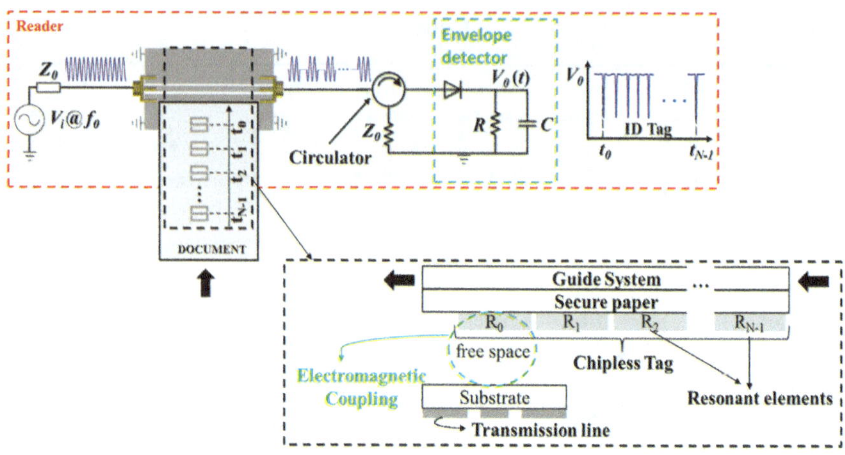

Fig. 2.2 Sketch showing the working principle of the near-field chipless-RFID system based on time-domain signature barcodes. The considered tag inclusions are S-SRRs

of a harmonic signal, properly tuned (plus the necessary electronics for tag reading). To read the tags sequentially (i.e., bit by bit) through near field, a transmission line fed by the interrogation signal and a mechanical system to guide the tag over the reader, in close proximity to it, are used [1, 6–12]. By this means, the time-varying electromagnetic coupling between the line and the functional resonant elements of the tag, resulting by tag motion, modulates the transmission coefficient at the frequency of the interrogation signal, and the signal at the output port of the line is AM modulated, being the ID code contained in the envelope function. Such envelope function, and consequently the ID code, can be inferred by means of a simple envelope detector.[3]

The working principle of these time-domain near-field chipless-RFID systems is illustrated in Fig. 2.2. It should be highlighted that in order to obtain the ID code, the distance between the tag and the reader (usually designated as air gap) must be small and as much uniform as possible. The maximum air gap is dictated by the maximum distance between the line and the tag that ensures an appreciable electromagnetic coupling (necessary to modulate the transmission coefficient of the line) when a functional tag resonator is on top of the line. Obviously, such distance depends on the topology and dimensions of the considered tag resonators, as well as on the reader line (type and size), and it is situated in the submillimeter range for the different chipless-RFID systems considered along this book. Air gap uniformity during tag motion (within certain limits) must be guaranteed in order to avoid overgaps, which may produce false readings. Nevertheless, as will be shown later, the

---

[3]Note that the output signal of the envelope detector is the time-domain signature of the tag, which can be visualized in an oscilloscope and typically exhibits peaks, or dips, corresponding to the different logic states ('1' or '0') of the resonant elements. The ID code can be also inferred by means of a digital post-processing unit.

proposed systems are quite tolerant against air gap variations.[4] It is also important to ensure a good lateral alignment between the tag and the reader for the same reasons, despite the fact that certain misalignments, unavoidable in a real system, can be tolerated. As it can be seen in the sketch of Fig. 2.2, an isolator is cascaded between the output port of the line and the envelope detector. Such element is used to protect the line against mismatching reflections caused by the diode, a highly nonlinear device. Alternatively, a matching network, usually incorporated in commercially available envelope detectors, can be considered for that purpose.

Although the data storage capacity in these near-field time-domain chipless-RFID systems is only limited by tag size, and tag reading proceeds by generating a simple harmonic interrogation signal, a mechanical guiding system for tag motion over the reader is needed as an essential part of the reader, as anticipated before. Such mechanical system must guarantee proper alignment between the tag and the reader and air gap uniformity (within certain limits, as discussed before). However, the presence of such mechanical system inherent to the reader does not represent a significant burden in terms of reader costs, since the required accuracy and robustness of the system for secure tag reading are comparable to those in commercially available products that use similar guiding systems (e.g., bank note readers). Nevertheless, it should be pointed out that the chipless-RFID systems based on time-domain signature barcodes, subject of the present book, may not be useful in applications requiring larger read distances and/or arbitrary orientation between the tag and the reader. At this point, it can be anticipated that the chipless-RFID systems discussed in this book are of special interest in applications where the read distance can be sacrificed in favor of data capacity. In this regard, the most canonical (but not unique) application is secure paper, where the ID codes (tags) can be directly printed on the document of interest (e.g., an official or corporate document, certificate, ballot, etc.), and used to provide security against counterfeiting or plagiarism. In this application, to be discussed in more detail later, tag reading by proximity (through near-field) may provide certain confidence of the system against spying or eavesdropping, as compared to other chipless-RFID systems based on far-field reading.

2.2 Strategies to Enhance the Data Density

A figure of merit in chipless-RFID systems is the data density of the tags. In frequency-domain or hybrid tags, both the DPF and the DPS have been the challenging parameters to be optimized (enhanced). In the chipless-RFID systems reported

---

[4]An excessive air gap may prevent from detecting functional resonators in the tag, thereby providing the logic state '0' in the corresponding resonant elements, regardless of their functionality. Conversely, if the gap is excessively small as compared to the nominal value (i.e., the optimum air gap used for system design), tag reading provides correct ID codes. Indeed, air gap oscillations during tag reading (unavoidable in practice) produce variations in the modulation index of the envelope function that can be tolerated provided such oscillations are constrained within certain limits.

in this chapter, where all the resonant elements, or inclusions, of the tags are identical, the DPF is virtually infinite and its consideration is out of interest. As for the DPS, obviously, achieving a high value of this parameter is fundamental in order to implement small-sized tags, given a certain number of bits (dictated by the considered application). However, as far as time-domain signature chipless-RFID tags are typically implemented as linear chains of resonant elements, or inclusions, their shape factor (defined as the ratio between the tag length and tag width) is usually very extreme (it is proportional to the number of bits). Therefore, the DPL, rather than the DPS, may be considered to be, in general, the most relevant parameter in tags based on chains of resonant elements, or inclusions [9, 13–15].

Note that optimization of the DPL does not necessarily drive to the optimization of the DPS. This may occur if very short and wide resonant elements, or inclusions, are used [13–15]. In this case, short tags (i.e., with high DPL) with tag width dictated by the width of the resonator (or inclusions), and hence with a DPS not necessarily small, are expected.[5] However, once the resonant element, or inclusion, of the tag is determined (i.e., its topology, dimensions and orientation with regard to the chain axis), DPL optimization, necessarily achievable by minimizing the separation between adjacent resonators, also enhances the DPS. Let us gain insight on this latter aspect. Typical resonant elements mostly used for the design of time-domain signature barcodes exhibit an equilibrated shape factor, with a square or rectangular geometry [1, 6–12]. The reason is that these geometries favor resonator excitation either through the magnetic or electric field generated by the reader line. With these resonant elements, good robustness of the chipless-RFID systems against misalignments and air gap variations has been demonstrated [10], and, for this reason, such resonators have been mostly imposed in the design of chipless-RFID systems based on time-domain signature barcodes [1, 6–12]. Evidently, the data density (both the DPS and the DPL) can be enhanced by reducing the size of such square or rectangular resonant elements, but this increases the frequency of the interrogation signal, which may raise reader costs, as well. Hence, the frequency of the interrogation signal cannot be considered to be a design parameter. Within this framework, and considering that the interrogation signal frequency is a design specification, it is clear that DPL (and DPS) optimization is achieved by reducing the inter-resonator distance to the limits imposed by the fabrication technology. There is a scarce margin for DPL optimization by tailoring the resonator shape factor, since extreme shape factors, corresponding to elongated rectangular geometries, usually degrade line-to-resonator coupling.[6]

Nevertheless, there are several types of resonant elements (driven by magnetic and/or electric field) which can be considered to be good candidates for tag design. Figure 2.3 depicts the typical topology (and the lumped element equivalent circuit

---

[5]In Sect. 2.3.3, it will be shown that by using linear (straight) resonant elements, or inclusions, oriented orthogonally to the chain axis, the DPL can be substantially enhanced, but in some cases at the expense of a degraded DPS, as compared to other time-domain signature chipless-RFID tags.

[6]Nevertheless, the tags based on linear strips, presented in Sect. 2.3.3.2, are an exception, as far as such inclusions (strips) behave as resonant elements excited by the quarter-wavelength stub resonator of the sensitive part of the reader. By contrast, the linear strips of the tags of Sects. 2.3.3.1 and 2.3.3.3 do not actually behave as resonant elements.

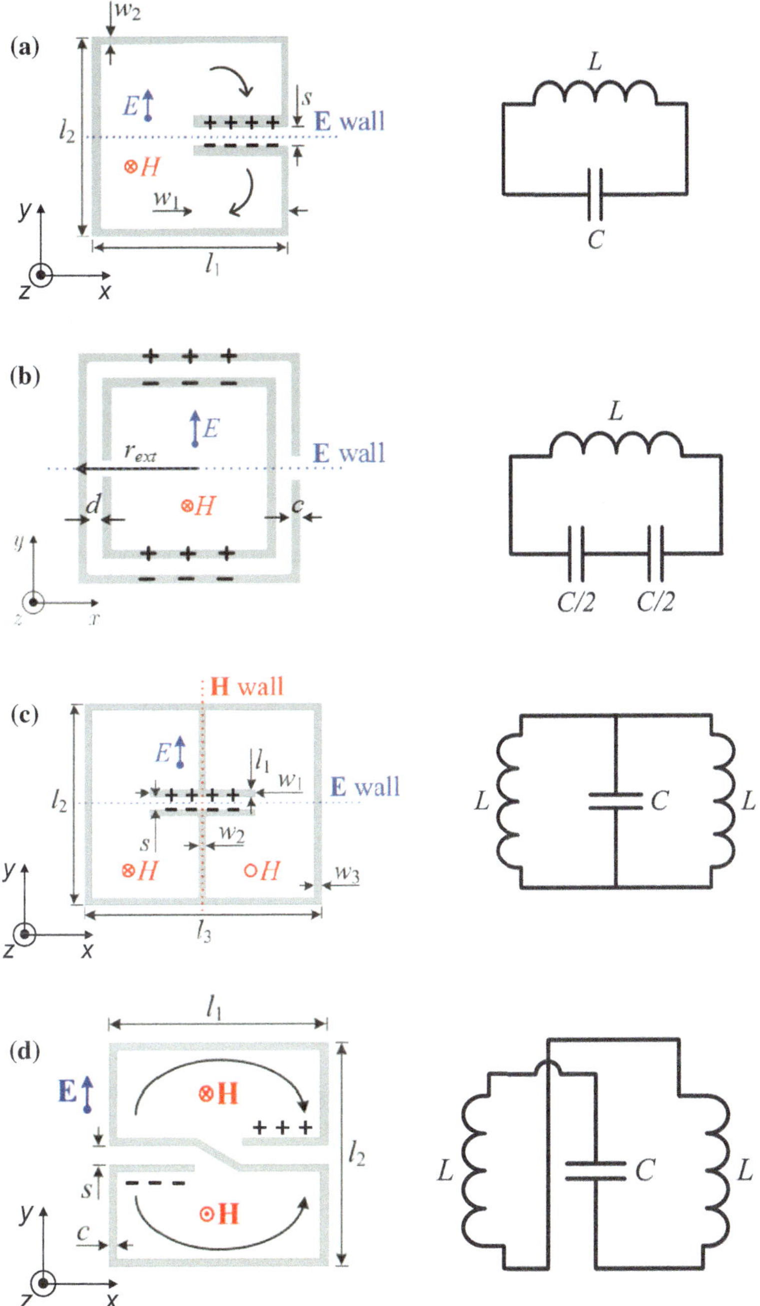

Fig. 2.3 Topologies and circuit models of several electrically small planar resonators, which can be useful for chipless-RFID tag implementation. **a** SRR, **b** EC-SRR, **c** ELC, **d** S-SRR. The direction of the electric and magnetic fields needed to excite the resonators, as well as the electric and magnetic walls (E-wall and H-wall, respectively) of the symmetric particles, are indicated

model) of such resonant elements, where a square geometry in all the cases has been considered. All these resonators are electrically small, and the elements of their circuit models can be inferred from a quasi-static analysis [16–18], out of the scope of this book. Two aspects are important to decide the preferred choice of resonant element for tag design: (i) the electrical size, and (ii) the specific driving mechanism. Concerning this latter aspect, the electric-LC (ELC) resonator [19, 20] and the S-SRR [21–24] can be excited at the fundamental resonance (the one of interest) by counter-magnetic fields applied in the two loops of the particles. The reason is that at the fundamental resonance, the currents in both loops flow in opposite directions [20, 24]. By contrast, the SRR [25] and the edge-coupled split ring resonator (EC-SRR) [17, 26, 27] need a uniform (or quasi-uniform) time-varying magnetic field for particle excitation at the fundamental resonance [17]. All these resonators can be also excited by means of a uniform (or quasi-uniform) time-varying electric field applied in the plane of the particle, in the direction indicated in Fig. 2.3 [17]. In view of these resonators' topologies and driving mechanisms, it follows that the ELC resonator and the S-SRR are good candidates to be used in tag design. The reason is that when the typical transmission lines, microstrip or CPW, are used in the reader for resonator excitation, the maximum coupling between the line and the resonator (corresponding to minimum transmission) is achieved when such resonant elements are on top of the line and perfectly aligned with it. Note that for ELC excitation, the particle must be oriented with its magnetic wall (indicated in Fig. 2.3) aligned with the line axis (also a magnetic wall for the fundamental quasi-microstrip and CPW modes of the microstrip and CPW line, respectively) [20]. Similarly, the S-SRR must be oriented with one loop present at one side of the line axis and with the other loop present at the other side [24].[7] By contrast, if SRRs or EC-SRRs are used for tag design, line to resonator coupling is cancelled when such resonators are perfectly aligned with the line axis (i.e., when the line axis is aligned with the symmetry plane of the particles) [28]. Under these conditions, particle excitation is prevented, and the injected signals are totally transmitted (except by the presence of losses) to the output port. With these resonant elements (SRRs and EC-SRRs), misalignment between the particle and the line is needed for line-to-resonator coupling. This situation must be avoided because there are two equivalent resonator positions, at both sides of the line axis, providing exactly the same frequency response of the line, and this is not convenient to detect the functional resonators in a reading operation.

Figure 2.4 depicts the simulated transmission coefficient of a CPW transmission line loaded with the considered resonant elements, in each case adequately positioned to achieve maximum line-to-resonator coupling. Note that for the SRR and the EC-SRR the maximum coupling, or equivalently, the maximum notch depth, takes place when the symmetry plane of such particles is aligned with the slots of the CPW transmission line. In view of this figure, it can be concluded that the electrically smaller particle is the S-SRR, as far as it provides the lowest resonance (or notch) frequency (note that particle dimensions and the considered substrates are identical

---

[7]Note that contrary to the ELC resonator, a bisymmetric particle, the S-SRR does not exhibit any symmetry plane. However, S-SRR excitation is very similar to ELC excitation.

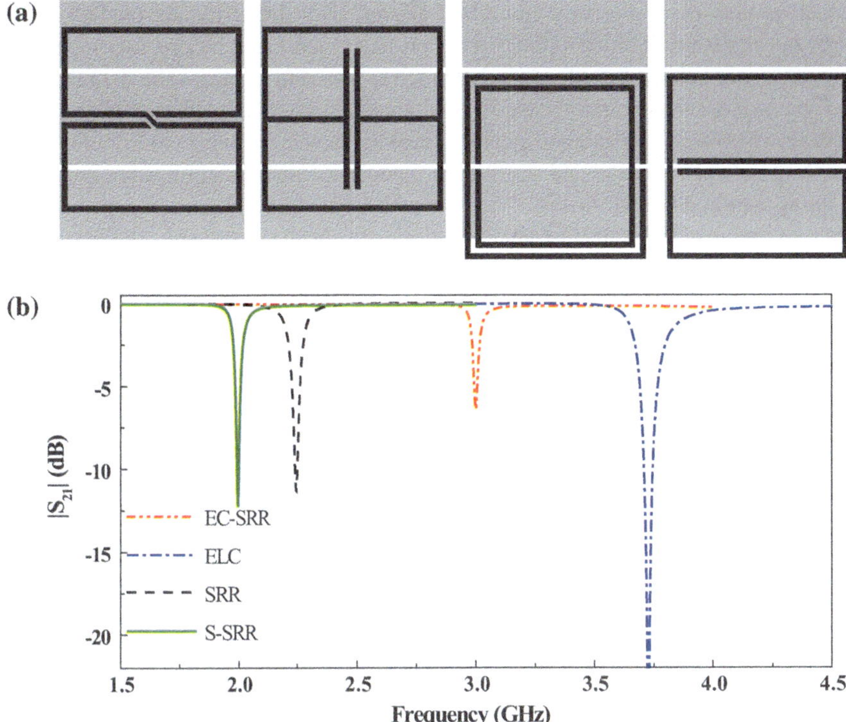

Fig. 2.4 Layout (**a**) and simulated transmission coefficient (**b**) of a CPW transmission line loaded with the resonant elements of Fig. 2.3, etched in the back substrate side (from left to right, S-SRR, ELC, EC-SRR and SRR). The simulation has been carried out by means of *Keysight Momentum*, where losses have been excluded. Substrate parameters for the CPW line are: dielectric constant $\varepsilon_r = 3.55$, and thickness $h = 0.81$ mm. The resonant elements are etched in the backside of the CPW. The slot and central strip width of the CPW are $G = 0.2$ mm and $W = 3.3$ mm, respectively. Resonator dimensions are (in mm): For the S-SRR: $l_1 = l_2 = 6.7$ and $c = s = 0.2$; For the ELC: $l_1 = 0.2$, $l_2 = l_3 = 6.7$, $w_1 = 5.1$ and $w_2 = w_3 = s = 0.2$; For the EC-SRR: $l_1 = l_2 = 6.7$ and $c = s = 0.2$; For the SRR: $l_1 = l_2 = 6.7$, $w_1 = 6.3$ and $w_2 = s = 0.2$

in all cases). Consequently, for a given reader (interrogation) frequency, resonator (and hence tag) size is expected to be minimized by considering the topology of the S-SRR. In summary, the combination of S-SRRs, printed or etched in the tag substrate, and a CPW transmission line, used as active part of the reader, seems to be a good solution for the implementation of chipless-RFID systems based on time-domain signature barcodes, as far as small-sized tags, easily readable with a simple fed line (a CPW), are expected. This solution will be discussed in detail in the next section, but additional approaches, using different tag resonators and reader lines, will be also studied in such section (tags based on non-resonant linear strips will be discussed, as well).

Let us now discuss a very important aspect, with direct impact on DPL (and DPS) optimization: multi-coupling. As it has been mentioned before, tiny separated resonators are needed in order to reduce the tag size and optimize the data density. However, under these circumstances, two negative effects, which may obscure tag reading, may arise: (i) inter-resonator coupling in the tag, and (ii) simultaneous coupling between the line and several tag resonators. Such couplings are indicated in the illustration shown in Fig. 2.5a, where a CPW (reader) loaded with S-SRRs (tag), etched in a different (movable) substrate, has been considered. The frequency response (transmission coefficient) that results when one of the S-SRRs of the tag is perfectly aligned with the line is depicted in Fig. 2.5b. The multi-notched response is indicative of the presence of multiple couplings, occurring simultaneously. With such many notches in the frequency response, it is not possible to unambiguously identify the functional or inoperative (either absent or detuned) resonators in the tags, and hence it is not possible to guarantee a correct tag reading. Note that this multi-coupling effect can be minimized, or even cancelled, by sufficiently spacing the resonant elements, but this goes against DPL and DPS optimization. Consequently, increasing inter-resonator distance is not a good solution.

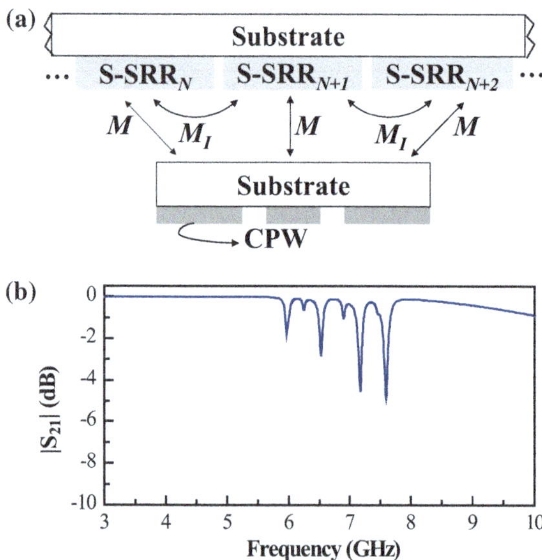

Fig. 2.5 **a** Sketch of the CPW (reader) loaded with an S-SRR-based tag with indication of the multiple couplings, including inter-resonator coupling (M_I) and line-to-resonator coupling (M). **b** Frequency response. The air gap is 0.25 mm. S-SRR dimensions, in reference to Fig. 2.3d are (in mm): $l_1 = 3.8$, $l_2 = 2.96$, $c = 0.4$ and $s = 0.2$. CPW dimensions are (in mm): slot width $G = 0.48$ and central strip width $W = 1.2$. The considered substrate for the CPW is the *Rogers RO3010* with dielectric constant $\varepsilon_r = 10.2$ and thickness $h = 635\ \mu m$. The considered substrate for the tag is *Rogers RO4003C* with dielectric constant $\varepsilon_r = 3.55$ and thickness $h = 203\ \mu m$

Fig. 2.6 **a** Sketch of the S-SRR-loaded CPW (reader) loaded with an S-SRR-based tag. **b** Frequency response. The dimensions and substrate parameters are those indicated in the caption of Fig. 2.5

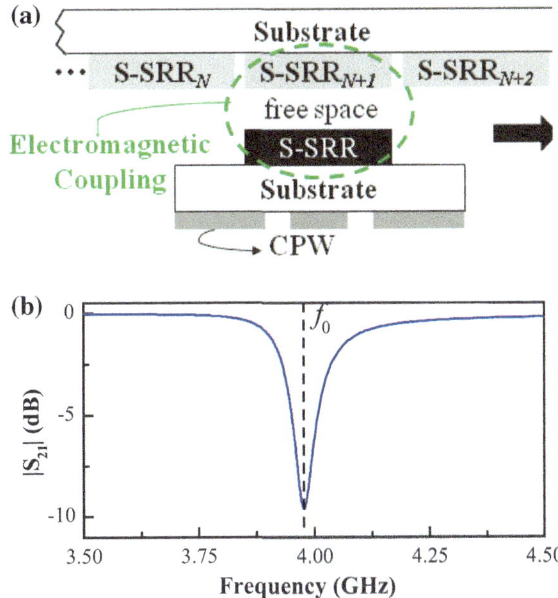

Let us now consider that the CPW transmission line of the reader is loaded with a resonator identical to those of the tag (S-SRRs) but rotated 180°, etched in the back substrate side and perfectly aligned with the line axis (Fig. 2.6a) [29]. When a resonant element of the tag is perfectly aligned with the (oppositely oriented) resonator of the line, both resonators can be considered to form a pair of coupled resonators, or a single resonant particle exhibiting two resonances, one below and the other one above the individual resonances of the isolated resonators.[8] In the case of study, where S-SRRs are considered, the resulting particle has been designated as broadside coupled S-SRR (BC-S-SRR) [6, 7] see Fig. 2.7. According to these words, the first resonance of the perfectly aligned S-SRRs (the one of the tag and the one of the line), or, equivalently, of the BC-S-SRR, is expected to lie substantially below the first resonance frequency of the isolated S-SRRs of the tag. Therefore, the aforementioned multi-coupling effect, manifested as a multi-notched response, is expected to be avoided by including an S-SRR in the reader line, as well as by tuning the harmonic feeding signal at the first resonance of the BC-S-SRR. Figure 2.6b depicts

---

[8]Like isolated S-SRRs, a pair of coupled S-SRRs exhibits multiple resonance frequencies. Such frequencies can be grouped in pairs, where the lower frequency of the pair lies below the resonance frequency/ies of the isolated resonators, and the higher frequency of the pair lies above the resonance frequency/ies of the isolated resonators. If both resonant elements are etched or printed on an identical substrate, the resonance frequencies of both isolated resonators are identical, and resonator coupling generates frequency splitting (i.e., a pair of resonances in the vicinity of the original resonances, separated a frequency that increases with the coupling level). If both resonant elements are not etched or printed on the same substrate, the resonances of the isolated particles do not coincide. Nevertheless, the effect of coupling is a further separation between both resonance frequencies.

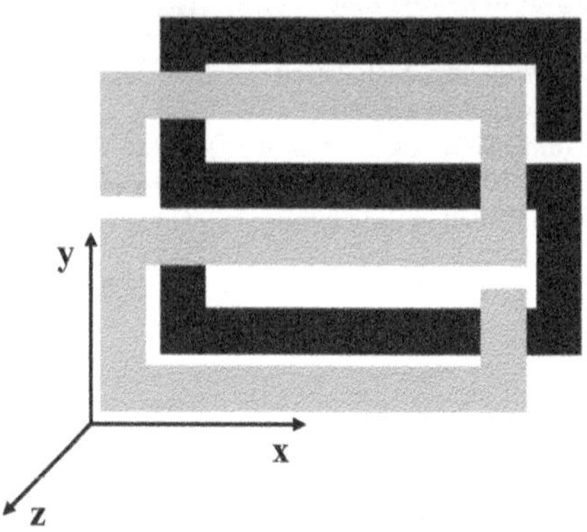

Fig. 2.7 Three-dimensional view of a BC-S-SRR

the frequency response of the structure of Fig. 2.6a, where it can be appreciated that a single notch appears, and that it is significantly shifted down, as compared to the frequencies of the multi-notches of Fig. 2.5b. Thus, the fact that the S-SRR-loaded reader with a tag resonator perfectly aligned with the S-SRR of the reader exhibits a frequency response with a single notch (Fig. 2.6b), opens the possibility to correctly read the tag ID code, yet keeping the inter-resonator distance to small values. As it will be latter shown, by tuning the frequency of the interrogation signal to the resonance frequency of the BC-S-SRR, f_0 (Fig. 2.6b), the transmission coefficient of the S-SRR-loaded line is expected to vary between a maximum value, close to 0 dB, when the resonant element of the tag is inoperative (absent or detuned), corresponding to the '0' logic state, and a minimum value, close to -10 dB in the case under consideration, when the S-SRR of the tag is functional (state '1') and it is perfectly aligned with the S-SRR of the reader. The ID code is, therefore, determined from the identification of minima in the transmission coefficient at f_0 when the tag is displaced above the reader, or equivalently, from the dips in the envelope function present at the output port of the line, the actual (and practical) reading mechanism. Note that such tag reading operation cannot be carried out with the reader line shown in Fig. 2.5a.

2.3 Examples of Time-Domain Signature Chipless-RFID Systems

In this section, several implementations of time-domain signature chipless-RFID systems are presented, analyzed, and compared. The working principle is the same in all the cases (see Sect. 2.1), but the configuration and type of reader line, as well as the resonant elements, or inclusions, of the tags, vary in the different implementations.

2.3.1 Band-Stop Configuration (S-SRR-Loaded CPW Reader)

The configuration reported in this section is, indeed, the one corresponding to the case study of Sect. 2.2, based on an S-SRR-loaded CPW transmission line as fundamental (sensitive) part of the reader, and tags consisting of chains of S-SRRs (rotated 180° with regard to the S-SRR of the line) [6, 7]. The designation of this configuration (band-stop) obeys to the fact that the S-SRR-loaded line of the reader exhibits a notched (i.e., band-stop) response.

2.3.1.1 Reader and Tag Design and Fabrication

For the design of the tag and the reader (S-SRR-loaded CPW), the first step is to set the frequency of the interrogation (harmonic) signal, or carrier signal,[9] f_c, to a reasonable value. In [6], where the band-stop configuration was reported for the first time, such frequency was set to $f_c = 4$ GHz. This frequency represents a tradeoff between the need to reduce the size of the S-SRRs of the tag (intimately related to f_c) as much as possible, and the implementation of low-cost readers. The complexity in the generation of harmonic signals increases with frequency. Moreover, by increasing the frequency of the interrogation signal, not only the size of the S-SRR of tag and reader is reduced, but also the maximum air gap distance and lateral misalignments compatible with safe readings are reduced, increasing the complexity and cost of the reader from a mechanical viewpoint. Therefore, it is clear that extremely reducing the tag size at the expense of an excessive increase of the interrogation signal frequency is not a good solution.

According to the value of f_c, the tag and reader line must be designed in order to exhibit the resonance f_0 at roughly f_c, where f_0 is the resonance frequency of the S-SRR-loaded line with a tag S-SRR perfectly aligned with the S-SRR of the line (let us call from now on REF position to such relative location between the tag and the reader line, illustrated in Fig. 2.6a, which provides the notch frequency at f_0). By this means, when a functional S-SRR lies on top of the S-SRR of the reader line, it will be detected by a minimum in the envelope function, provided $f_c \approx f_0$. Contrarily, if the S-SRR is not present or it is detuned, it is expected that the amplitude of the output signal is similar to the amplitude of the feeding signal.

Another important aspect in the design of the tag and the reader line is the air gap distance, since this has direct influence on f_0, the first resonance frequency of the BC-S-SRR pair (see Sect. 2.2). Therefore, a nominal air gap value must be considered for design purposes. In [6, 7], such air gap distance was set to 0.25 mm, which is a reasonable value taking into account the potential mechanical vibrations in a practical guiding system.

---

[9]The interrogation signal is AM modulated by tag motion at the output port of the reader line, as discussed in Sect. 2.1. Therefore, such signal can be also designated as carrier signal, and the corresponding frequency as f_c.

The considered substrate for the implementation of the S-SRR-loaded reader line is the *Rogers RO3010* with dielectric constant $\varepsilon_r = 10.2$ and thickness $h = 635$ μm. The CPW slot and central strip widths are $G = 0.48$ mm and $W = 1.2$ mm, respectively, corresponding to a 50-Ω line impedance. For the design of the S-SRR, we have chosen a rectangular geometry with the longer side parallel oriented to the line axis (orthogonal to the tag chain axis). The width of the S-SRR has been set to $c = 0.4$ mm, a relatively wide value that provides an acceptable notch depth and width. Guidelines for S-SRR design, out of the scope of this book, can be found in [30]. However, the notch depth and width increase as c increases, since the S-SRR inductance and resistance decrease, whilst the capacitance increases. Once c is set to a certain value (0.4 mm in the case study), the single free parameter to tune the resonance frequency f_0 is the length of the loops, or, alternatively, the side lengths, l_1 and l_2. The layout of the S-SRR-loaded CPW with indication of the relevant dimensions is depicted in Fig. 2.8. The S-SRR slot has small influence on particle behavior, and it has been set to $s = 0.2$ mm. It should be pointed out that dimensions l_1 and l_2, given in the caption of Fig. 2.8, have been determined in order to set the first resonance of the BC-S-SRR (by considering an air gap of 0.25 mm) to $f_0 = 4$ GHz, the frequency of the interrogation signal, f_c. The considered substrate for tag fabrication is the *Rogers RO4003C* with dielectric constant $\varepsilon_r = 3.55$ and thickness $h = 203$ μm (this has been chosen as far as these substrate parameters are close to the typical parameters encountered in flexible plastic or paper substrates). However, with such narrow tag substrate, attachment of the tag to a more rigid substrate (*FR4* with dielectric constant $\varepsilon_r = 4.7$ and thickness $h = 1.6$ mm in our case) is needed, in order to provide mechanical stability [6, 7]. This means that for the determination of l_1 and l_2, providing the required value of f_0, both dielectric layers in the tag side, plus the air gap and reader substrate have to be taken into account. The final values of such geometric parameters are $l_1 = 3.8$ mm and $l_2 = 2.96$ mm. Finally, the separation between adjacent S-SRRs in the tag chain is 0.2 mm, corresponding to a chain period of 3.16 mm. The photograph of the first version of this reader line and tag (with 10 S-SRRs, equivalent to 10 bits) is depicted in Fig. 2.9 [6], where a guiding channel can be seen in the back side of the CPW to ease tag motion.

The frequency responses of the S-SRR-loaded CPW of the reader loaded with tags placed at different relative positions are depicted in Fig. 2.10. It is apparent from this figure that by tuning the carrier frequency to $f_c = f_0$, the excursion experienced by the transmission coefficient at that frequency, due to the effects of tag motion, is expected

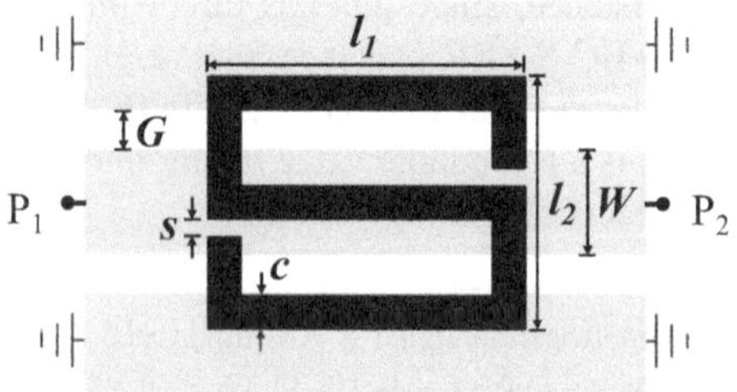

Fig. 2.8 Layout of the S-SRR-loaded CPW and relevant dimensions. CPW dimensions are (in mm) $W = 1.2$ and $G = 0.48$; S-SRR dimensions are (in mm) $l_1 = 3.8$, $l_2 = 2.96$, $c = 0.4$ and $s = 0.2$

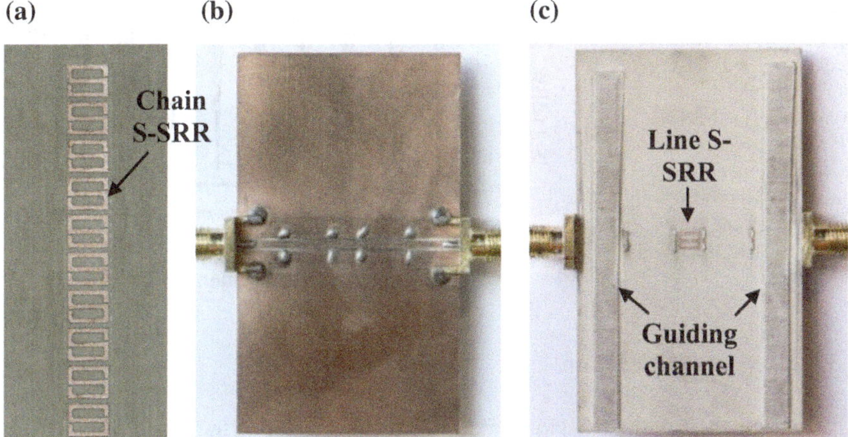

Fig. 2.9 Photograph of the fabricated 10-S-SRR (10-bit) tag (**a**), top (**b**), and bottom (**c**) views of the S-SRR-loaded CPW transmission line of the reader. The photographs do not correspond to the same scale

Fig. 2.10 Simulated (**a**) and measured (**b**) frequency responses of the S-SRR-loaded CPW loaded with an S-SRR tag with the chain located at different relative positions. The simulations have been inferred from *Keysight Momentum*, whereas the measured responses have been obtained by means of the *Agilent N5221A* vector network analyzer. Reprinted with permission from [7]; copyright 2017 IEEE

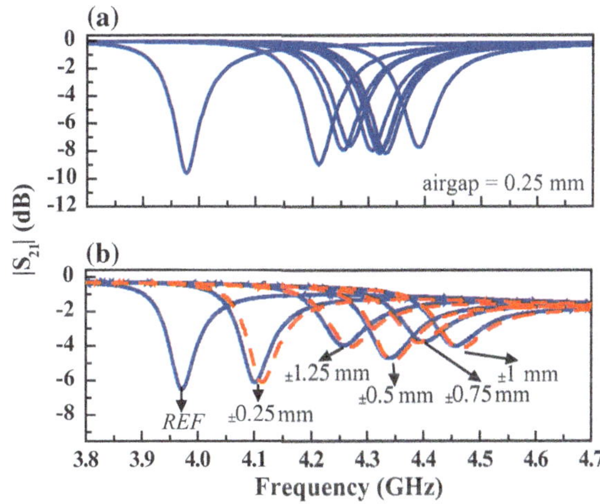

to be significant. Indeed, it can be appreciated that for a ± 0.25 mm displacement from the REF position, the insertion loss at f_0 is negligible. Figure 2.11 depicts the variation of the transmission coefficient at f_0, where it can be seen that dips of roughly 8 dB depth (electromagnetic simulation) appear each time an S-SRR of the tag is perfectly aligned with the reader S-SRR (REF position).

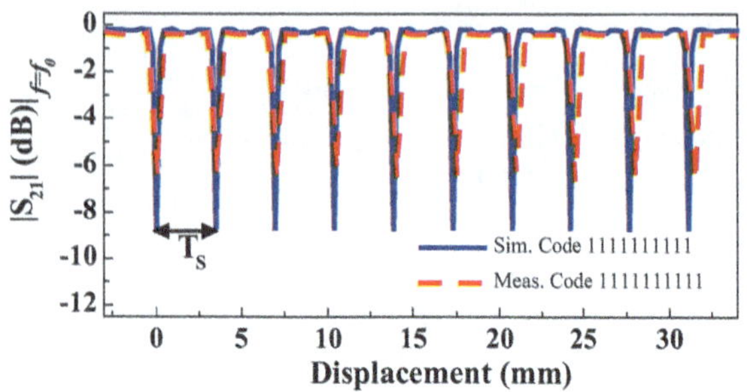

Fig. 2.11 Attenuation of the tag-CPW at the reference frequency, f_0, as the tag moves completely (transversally) above the CPW of the reader. All the resonant elements are functional. Reprinted with permission from [6]; copyright 2017 IEEE

2.3.1.2   System Validation

The experimental set-up used to validate tag reading, based on the scheme of Fig. 2.2, is depicted in Fig. 2.12. The carrier signal is generated by means of the *Agilent E4438C* function generator. The circulator employed to implement the isolator is the *ATM ATc4-8*. The envelope detector uses the *Avago HSMS-2860* Schottky diode and the *N2795A Active probe* (which acts as lowpass filter with $R = 1$ MΩ and $C = 1$ pF), connected to an oscilloscope (the *Agilent MSO-X-3104A*) in order to visualize the envelope function. The measured normalized envelope functions, corresponding to three different codes, are depicted in Fig. 2.13, where it can be appreciated that the '1' states provide dips in the reading windows, contrary to the '0' states, which have been generated by the absence of S-SRR in the corresponding position.

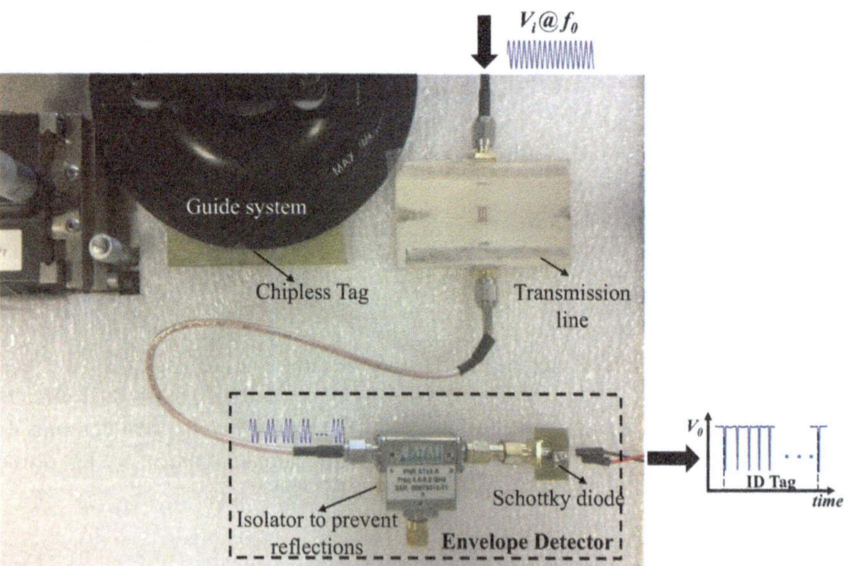

Fig. 2.12 Experimental set-up for system validation

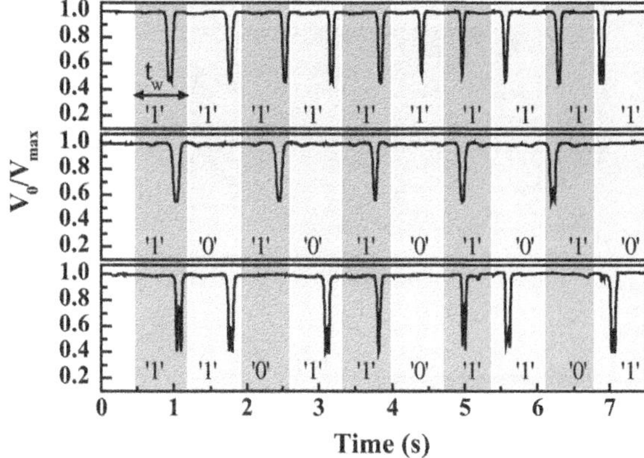

Fig. 2.13 Measured normalized envelope for 3 fabricated tags with the indicated codes. The dips are not exactly at same positions within the readout windows since the in-house guiding system does not guarantee uniform displacement velocity of the tag over the reader. Reprinted with permission from [6]; copyright 2017 IEEE

A second version of the tag and the reader, based on identical resonant elements and line, is reported next [7]. Actually, the unique difference concerns the tag chain, which is arranged forming a circle in this second prototype. This eases validation as far as the tag can be attached to a rotor, driven by a step motor, which provides a uniform tag velocity. Nevertheless, the S-SRR-loaded line has been slightly modified for better alignment with the rotor. The photographs of the fabricated tags and reader line are depicted in Fig. 2.14. Four different 40-bit tags, each one etched on a quadrant of the circled substrate, are visible in the figure (where, again, the absence of resonant element in a certain position in the chain is considered to generate the corresponding '0' logic state). The measured normalized envelope functions of the four tags, also inferred from the set-up shown in Fig. 2.12 (but using the *STM 23Q-3AN* stepper motor, see also Fig. 2.14) are depicted in Fig. 2.15.

In the tags of Figs. 2.9 and 2.14, the data density per surface and per length unit are DPS $= 7.4$ bit/cm^2 and DPL $= 3.16$ bit/cm, respectively. In order to improve the DPL, the relevant figure of merit in these chipless-RFID tags, as discussed before, one possibility is to consider a double chain of S-SRRs in the tag, with a relative displacement between chains of one semi-period [9]. With this tag topology, the reader needs to be modified by including two S-SRRs, rather than one, in the back side of the CPW transmission line. This new tag and reader line are depicted in Figs. 2.16 and 2.17, respectively. Figure 2.18 depicts the transmission coefficient at f_0 (inferred from electromagnetic simulation) obtained when a double-chain of S-SRRs is displaced over the reader. The 10 dips are indicative of the 5 S-SRRs present in each chain, corresponding to a 10-bits tag. Note that the period of the response is half the period of the response depicted in Fig. 2.11 because the density of S-SRRs per unit length is twice the one of the tags of Fig. 2.9. In the double-chain S-SRR tags, the data densities are DPL $= 5.7$ bit/cm and DPS $= 5.53$ bits/cm^2, i.e., such tags exhibit a substantially improved DPL, as compared to the one of the tags in Figs. 2.9 or 2.14. The validation has been carried out by considering the experimental set-up described before, and the reading results of three different tags are depicted in Fig. 2.19. In this case, however, resonator detuning (by cutting the resonators)

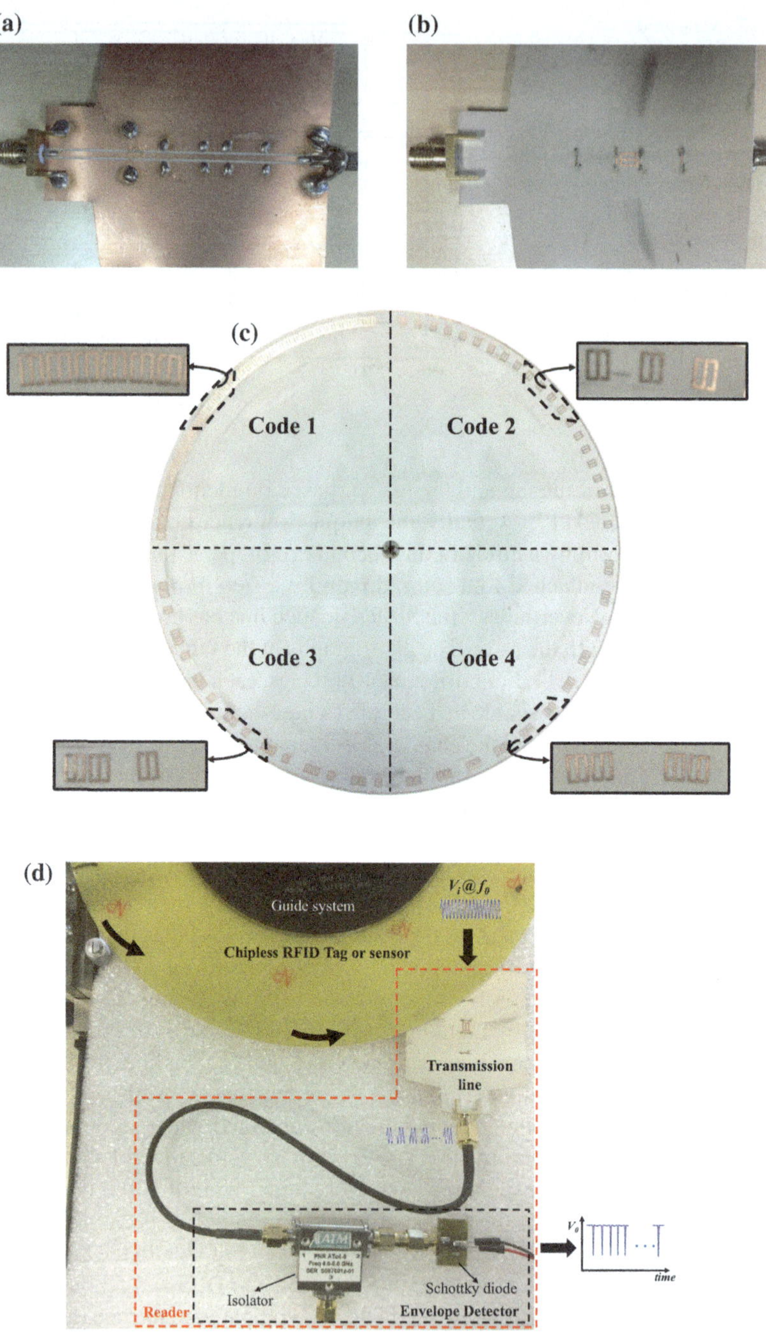

Fig. 2.14 Photograph of the transmission line reader, including CPW (**a**), the S-SRR (**b**), fabricated tags (**c**), and experimental set-up (**d**)

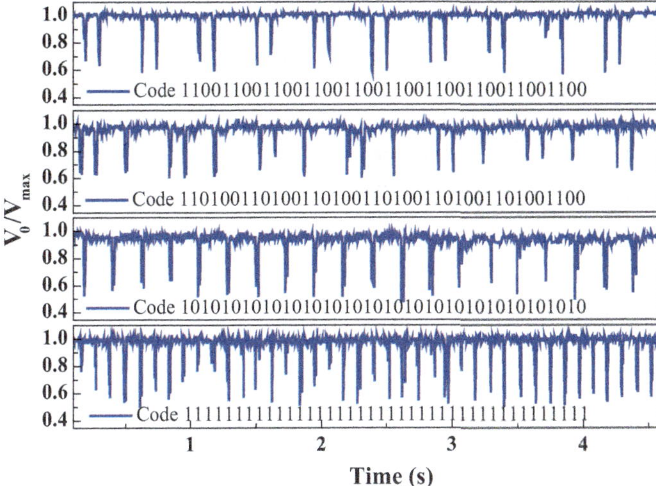

Fig. 2.15 Measured normalized envelope for the four 40-bit fabricated chipless tags with the indicated ID codes. Reprinted with permission from [7]; copyright 2017 IEEE

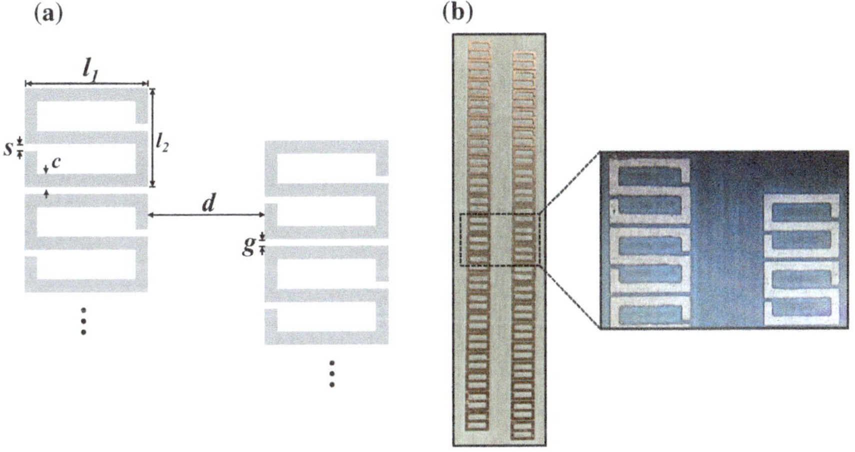

Fig. 2.16 Layout (zoom) of the double chain of S-SRRs (**a**) and photograph of the fabricated 40-bit chipless RFID tag (**b**). S-SRR dimensions are (in mm) $l_1 = 3.8$, $l_2 = 2.96$, $c = 0.4$, $s = 0.2$, $g = 0.2$ and $d = 3.6$. Reprinted with permission from [9]; copyright 2017 IEEE

has been the considered procedure to "write" the desired logic '0' states in the tag. It can be appreciated that the different codes are correctly read, which means that cutting the resonant elements is as efficient as eliminating them in order to make them inoperative (this aspect will be discussed in further detail in Chapter 3).

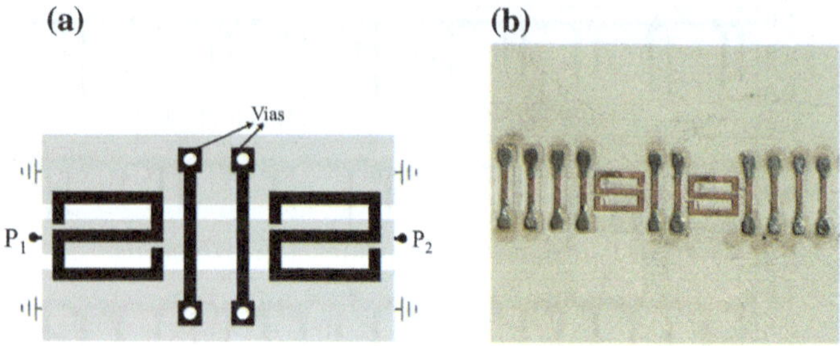

Fig. 2.17 Layout (**a**) and photograph (**b**) of the designed and fabricated S-SRR-loaded CPW acting as sensitive part of the reader. In (**b**) only the backside of the CPW is shown

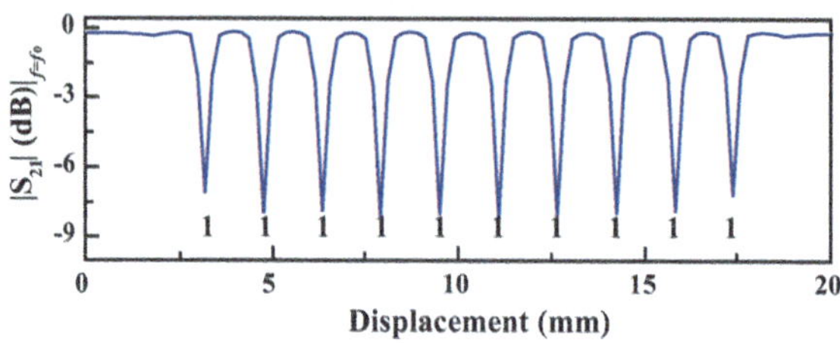

Fig. 2.18 Transmission coefficient at f_0, as the linear 10-bit tag is displaced above the S-SRR-loaded CPW transmission line, inferred from electromagnetic simulation (using *Keysight Momentum*), by considering an air gap of 0.25 mm. Reprinted with permission from [9]

Fig. 2.19 Envelope functions of the 40-bit tags with the indicated codes. Reprinted with permission from [9]

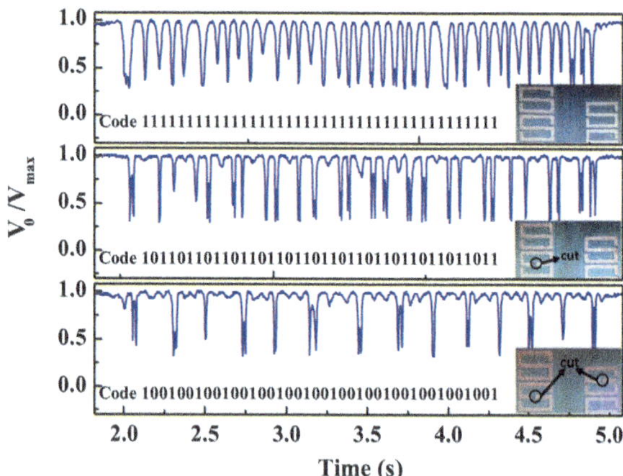

2.3.1.3   Tolerance Analysis

In practice, deviations from the nominal geometric values in the proposed chipless-RFID systems are unavoidable. Not only the dimensions of the S-SRRs, but also the air gap separation and the lateral alignment between the S-SRR of the line and the

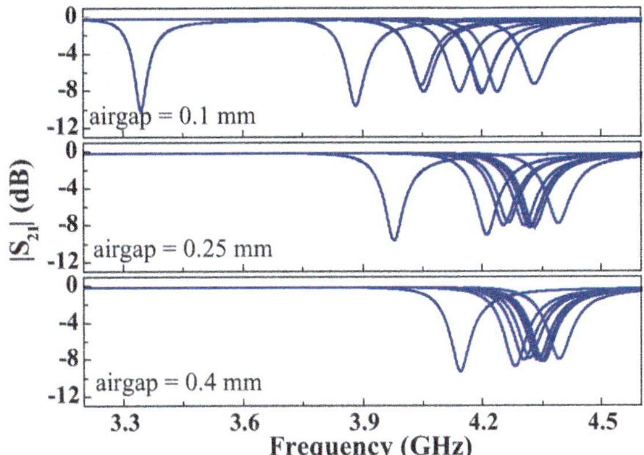

Fig. 2.20 Transmission coefficient, inferred from electromagnetic simulation using *Keysight Momentum*, for different relative displacements between the S-SRR chain and the CPW, parameterized by the air gap separation. Reprinted with permission from [7]; copyright 2017 IEEE

S-SRRs of the tags may be subjected to slight variations. With regard to the S-SRR geometry, current etching and printing systems provide a reasonable accuracy for the considered critical dimensions. Moreover, potential variations in the S-SRR or BC-S-SRR resonance frequency (the latter one, f_0, being a design parameter), caused by fabrication related tolerances, can be compensated by tuning the carrier frequency so that $f_c \approx f_0$.

The most critical parameter is the air gap distance, given that it determines the frequency of the perfectly aligned S-SRRs (or BC-S-SRR), f_0. Moreover, the air gap may vary during a reading operation due to mechanical instabilities (vibration, lack of flatness in the tag, etc.). The effects of the air gap variation can be appreciated from the simulation of the transmission coefficient for different displacements between the tag and the S-SRR-loaded line of the reader, by considering different air gaps below and above the nominal value. Such simulations (including the one of Fig. 2.10a, corresponding to the nominal air gap) are depicted in Fig. 2.20. By increasing the air gap, the resonance frequency of the BC-S-SRR increases, since the coupling between the individual S-SRRs decays. Such resonances are given by the frequencies of the first notch in each graph of Fig. 2.20. The notch situated at the highest frequency of each graph corresponds to a relative displacement of a half semi-period, and it barely changes with the air gap. The reason is that for a half semi-period displacement, the coupling between the S-SRR of the line and the S-SRRs of the tag is negligible, and the notch is roughly given by the resonance of the isolated S-SRR of the line. An excessive air gap must be avoided since the frequency difference between the extreme notches is small, and this may difficult the tuning of the carrier frequency. Given a nominal gap distance, such frequency must be tuned to f_0, as mentioned before. However, if it is tuned to the left of f_0 (e.g., due to the presence of an air gap separation slightly above the nominal one), reading may be obscured by the small or negligible excursion of the transmission coefficient at f_c, potentially providing false ID codes. On the contrary, if $f_c > f_0$ (gap separation below the nominal value), then two dips per S-SRR are expected, one for that position where the resonance of the pair of (misaligned) S-SRRs coincides with f_c, and the other one for the equivalent position at the other side of the line axis. This explains that in Figs. 2.13, 2.15 and 2.19 some dips are split. To avoid false readings, the gap has been deliberately under-dimensioned, but unavoidable variations in the air gap produce single dips for certain bits, with varying depth (an excess of air gap reduces the excursion of the transmission coefficient at f_0).

In view of Fig. 2.20, an air gap separation slightly above 0.25 mm (for f_c tuned to 4 GHz) can be tolerated, since the queue of the notch at f_0 of the corresponding gap may produce an appreciable variation of the transmission coefficient at f_c. Note, however, that an air gap of 0.4 mm is excessive since the transmission coefficient at f_c is roughly 0 dB (see Fig. 2.20).

Concerning the effects of lateral misalignments between the tag and the reader, Fig. 2.21a depicts the responses for different relative lateral displacements by considering that the S-SRR of the tag is perfectly aligned in the longitudinal direction with the S-SRR of the reader. The nominal air gap of 0.25 mm is considered. Figure 2.21b depicts the magnitude of the transmission coefficient at f_0 for the indicated lateral displacements. From this figure, it follows that the system is (roughly) ± 0.25 mm tolerant against lateral shifts, as far as the magnitude of the transmission coefficient within this range is better than -5 dB.

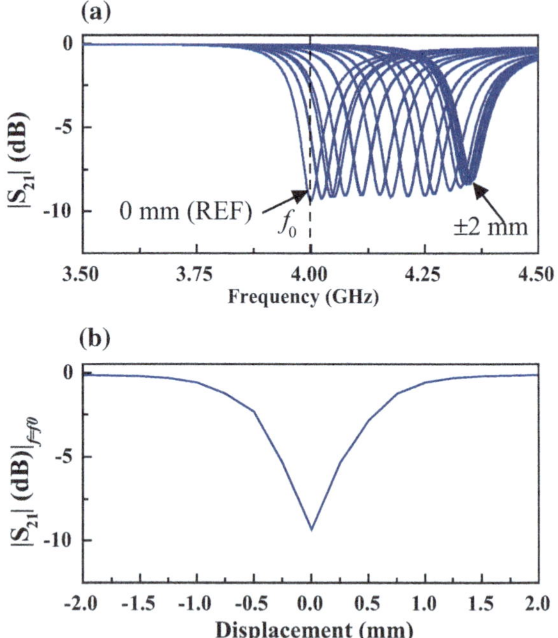

Fig. 2.21 Transmission coefficient, inferred from electromagnetic simulation using *Keysight Momentum*, for different relative lateral displacements between the S-SRR chain and the CPW (**a**) and magnitude of the transmission coefficient at f_0 as a function of the lateral displacement (**b**)

2.3.2 Band-Pass Configuration (SRR-Loaded Microstrip Line Reader)

In the band-stop configuration reported in the previous section, the output dynamic range, or excursion experienced by the transmission coefficient at f_c when an S-SRR of the tag is perfectly aligned with the S-SRR of the reader, is roughly 6 dB (see, e.g., Figure 2.11). This value is enough to discern the two binary states associated to the resonant elements of the tag, but the variation of the normalized envelope function is limited. More specifically, the depth of the dips associated to the logic states '1', is roughly 40–60% the value of the maxima, as can be appreciated in Figs. 2.13, 2.15 and 2.19. Increasing the dynamic range is convenient in order to provide major system robustness against noise or spurious effects, which may have direct impact on the bit-error-rate (BER). Particularly, to detect the '1' logic states by threshold value, optimization (reduction) of the ratio between the minima and the maxima in the normalized envelope function is necessary.

As an alternative to the chipless-RFID system reported in the Sect. 2.3.1, a band-pass configuration, with the reader line implemented in microstrip technology, was proposed in [10]. Besides the intrinsic advantages of microstrip lines over CPWs (essentially backside isolation and lack of parasitic modes[10]), the fundamental benefit of this near-field chipless-RFID system based on a bandpass microstrip structure is the high achievable dynamic range. Such improved dynamic range is a result of the transmission zero present in the frequency response of the considered structure, to be discussed next.

2.3.2.1 Reader and Tag Design

The reader line of the system consists of an order-1 microstrip band-pass filter implemented by means of a capacitively coupled square-shaped SRR, and the tag is a chain of SRRs identical to those of the reader, but rotated 180°. The layout of the reader line, including relevant dimensions, is depicted in Fig. 2.22a. In [10], the reader filter was implemented on the *Rogers RO3010* substrate with dielectric constant $\varepsilon_r = 10.2$, thickness $h = 0.635\ \mu$m, and loss tangent $\tan\delta = 0.0022$. Such structure, without tag cover, exhibits the simulated frequency response depicted in Fig. 2.23. The presence of a pole (where the signal is totally transmitted, except by the effect of losses) and a transmission zero can be appreciated. This frequency response is very useful due to the large insertion loss that it exhibits at the transmission zero frequency (>50 dB). Such high insertion loss has direct impact on the modulation index of the AM signal

---

[10]In CPW transmission lines, the parasitic slot (odd) mode may be generated if symmetry with regard to the line axis is disrupted. For that reason, in the reader lines of Figs. 2.9, 2.14 and 2.17, backside strips connecting the ground planes through vias are added. This prevents mode conversion from the fundamental (even) CPW mode to the slot mode, but at the expense of further fabrication complexity.

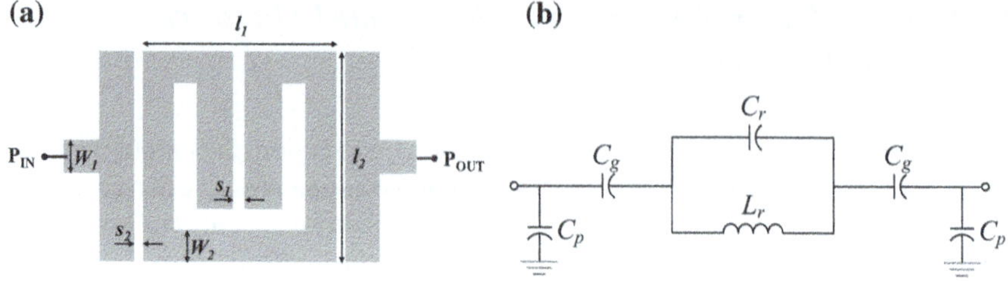

Fig. 2.22 Topology of the reader line and relevant dimensions (**a**) and lumped element equivalent circuit (**b**). Dimensions are (in mm): $l_1 = 3.16$, $l_2 = 3.35$, $s_1 = 0.2$, $s_2 = 0.2$, $W_1 = 0.56$, and $W_2 = 0.5$

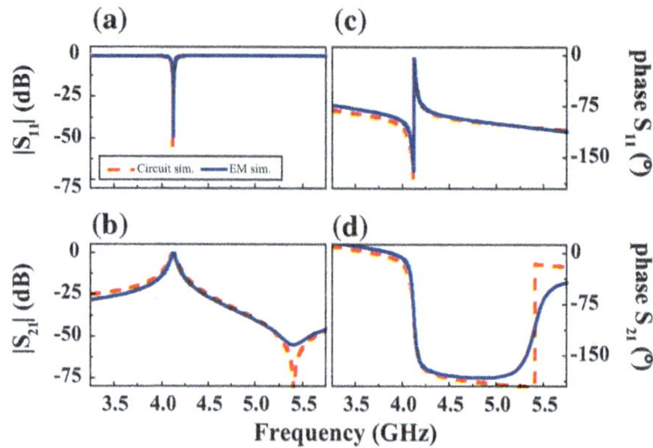

Fig. 2.23 Frequency response of the structure of Fig. 2.22 inferred from lossless electromagnetic and circuit simulation. **a** Magnitude of S_{11}. **b** Magnitude of S_{21}. **c** Phase of S_{11}. **d** Phase of S_{21}. The extracted parameters of the lumped element equivalent circuit (in reference to Fig. 2.22) are: $L_r = 27.1$ nH, $C_r = 0.03$ pF, $C_g = 0.05$ pF and $C_p = 0.76$ pF. Electromagnetic simulations have been carried out with the *Momentum* tool of the *Keysight ADS* commercial software, whereas circuit simulations have been carried out with the circuit simulator of that software. Reprinted with permission from [10]; copyright 2017 IEEE

at the output port of the reader line (and, consequently, on the ratio between the dip and peak values in the normalized envelope function, as will be later shown).

To gain insight on the design of the reader line filter, the lumped element equivalent circuit, which neglects losses as well as the effects of the access lines, is useful (Fig. 2.22b). The transmission zero frequency is given by the intrinsic resonance frequency of the SRR, i.e.,

$$f_z = \frac{1}{2\pi \sqrt{L_r C_r}} \tag{2.1}$$

where the series branch opens. The position of the pole is not given by a simple expression. At the pole frequency, f_r, the iterative impedance (or image impedance), given by

$$Z_0(\omega) = \sqrt{\frac{Z_s(\omega)Z_p(\omega)/2}{1 + \left(Z_s(\omega)/2Z_p(\omega)\right)}} \tag{2.2}$$

coincides with the reference impedance of the ports, typically 50 Ω (Z_s and Z_p are the impedances of the series and shunt branches, respectively). Thus, the element values must be chosen to satisfy expressions (2.1) and (2.2) with the desired values of f_z and f_r, and by forcing expression (2.2) to be 50 Ω at f_r. Nevertheless, note that (2.1) and (2.2) do not determine univocally the element values of the circuit model. For that purpose, two additional conditions are needed. The frequency where the series branch shorts, f_s, may be one of such conditions. This frequency is given by

$$f_s = \frac{1}{\pi\sqrt{2L_r\left(C_g + 2C_r\right)}} \tag{2.3}$$

This frequency can be inferred from the reflection coefficient, S_{11}, represented in the Smith chart, since at that frequency the trace of S_{11} intersects with the unit conductance circle. From the value of the parallel reactance, directly given by the Smith chart, the shunt capacitance value, C_p, is derived, this being the fourth condition.

Once the element values of the circuit model satisfying the system requirements (mainly, f_z and f_r) have been found, the generation of the layout is not straightforward. However, extracting the circuit elements from the layout (the inverse process) is very simple, as long as the relevant parameters (f_z, f_r, f_s and C_p) needed for that purpose, can be directly inferred from the electromagnetic simulation of the layout [18, 31]. Thus, with the help of the circuit model and parameter extraction, few iterations typically suffice to generate the layout adapted to the requirements of the system. The circuit simulation of the structure of Fig. 2.22, with extracted parameters indicated in the caption, is also included in Fig. 2.23. An excellent agreement, pointing out the validity of the model, can be appreciated (some discrepancies in the vicinity of f_z are due to radiation, not accounted for by the circuit model).

The frequency response depicted in Fig. 2.23 corresponds to the SRR-loaded microstrip line without tag loading, i.e., surrounded by air. Let us now consider that the structure is loaded with the tag, and particularly for a tag position corresponding to perfectly aligned SRRs (the distance between adjacent SRRs, if they are present, in the tag chain is 0.2 mm, corresponding to a chain period of 3.36 mm). For this position, referred to as reference position (REF), the pair of vertically aligned SRRs forms the so-called broadside-coupled SRR (BC-SRR) [16–18]. Similar to the BC-S-SRR, this composite particle is characterized by a strong electric coupling between the individual SRRs, provided the air gap separation between them is small enough. The consequence is a significant decrease of the fundamental resonance frequency of the composite particle (BC-SRR), resulting in an overall shift of the transmission coefficient towards lower frequencies. By considering that the air gap separation is 265 μm (this value will be justified later), that the SRR of the tag is etched on the *Rogers RO4003C* substrate with thickness $h = 0.204$ mm, dielectric constant $\varepsilon_r = 3.55$ and loss tangent tan$\delta = 0.0021$, and that this narrow substrate is attached to *FR4* (with thickness $h = 1.6$ mm, dielectric constant $\varepsilon_r = 4.7$ and loss tangent tan$\delta = 0.014$) for mechanical stability, the resulting frequency response (inferred from electromagnetic simulation) is the one depicted in Fig. 2.24. Note that if the tag is attached to a different material, its effects can be taken into account at simulation and design level. Nevertheless, some tolerance exists in the material used to provide mechanical stability, due to the large excursion of the transmission coefficient experienced by tag motion. The circuit simulation with extracted parameters is also depicted in Fig. 2.24 (the parameters are indicated in the caption). Again, good agreement between the lossless circuit and electromagnetic simulation is obtained. Interestingly, C_g and C_p have not experienced a significant variation as compared to the structure without tag cover, as expected. The main variation corresponds to the capacitance of the resonant element (BC-SRR), which has experienced an increase of roughly four times due to the broadside (face-to-face) capacitance between the metal strips of the particle.

The frequency response of the tag loaded structure for different relative positions between the SRR of the tag and the one of the line is depicted in Fig. 2.25 (tag displacement is in the direction of the line axis, and, without loss of generality, only one SRR in the tag has been considered). In this case, losses have been considered, since the maximum and minimum values of the transmission coefficient are important. The considered air gap separation is the one in reference to the response of Fig. 2.24 (i.e., 265 μm). Departure from the REF position modifies (shifts up) the transmission and reflection zero frequencies. Figure 2.26 depicts the variation of the transmission

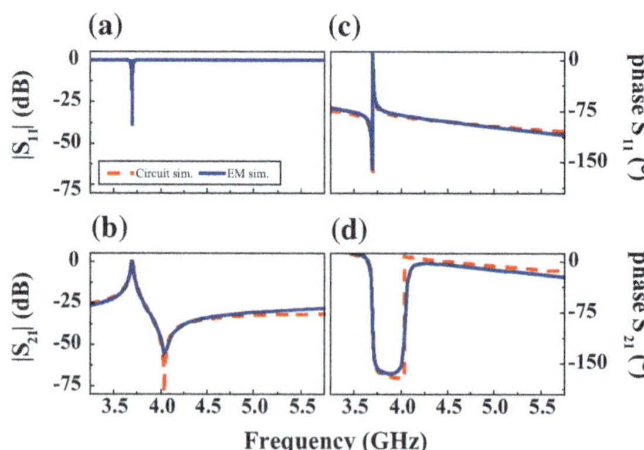

Fig. 2.24 Frequency response of the structure of Fig. 2.22 with tag cover as explained in the text, inferred from lossless electromagnetic and circuit simulation. **a** Magnitude of S_{11}. **b** magnitude of S_{21}. **c** phase of S_{11}. **d** phase of S_{21}. The extracted parameters of the lumped element equivalent circuit are: $L_r = 11.9$ nH, $C_r = 0.13$ pF, $C_g = 0.05$ pF and $C_p = 0.87$ pF. Reprinted with permission from [10]; copyright 2017 IEEE

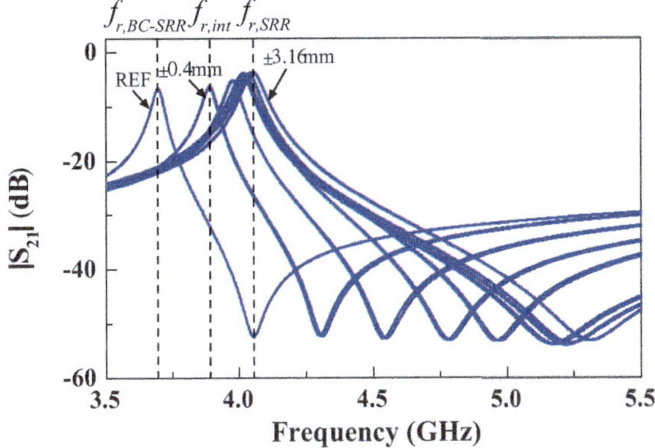

Fig. 2.25 Transmission coefficient (magnitude) of the SRR-loaded line of Fig. 2.22 with tag cover, for different relative positions between the SRR of the line and the SRR of the tag. Relevant frequencies, discussed in the text, are indicated. These results have been inferred by electromagnetic simulation. Reprinted with permission from [10]; copyright 2017 IEEE

Fig. 2.26 Variation of the transmission coefficient as a function of the relative displacement for the indicated frequencies. These results have been inferred by electromagnetic simulation. Reprinted with permission from [10]; copyright 2017 IEEE

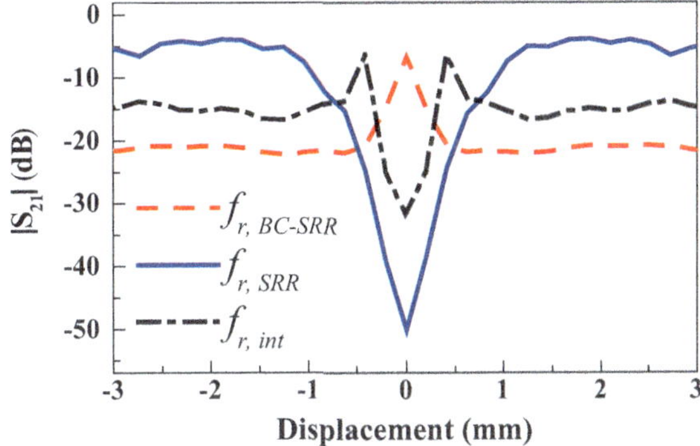

coefficient as a function of the relative displacement for specific frequencies (indicated in the figure). One of these frequencies is the reflection zero frequency for the case of perfectly aligned SRRs (forming the BC-SRR). Let us designate this frequency as $f_{r,BC\text{-}SRR}$. The transmission coefficient for this frequency is a maximum for the REF position. Departure from this position reduces the transmission coefficient, but it saturates to roughly -22 dB for a relatively small displacement. The variation of the transmission coefficient at the reflection zero frequency of the structure with completely misaligned SRRs (i.e., not overlapping) is also depicted in Fig. 2.26. Note that this frequency is similar, but not identical, to the reflection zero frequency of the structure without tag loading. The reason is that the presence of the tag substrate slightly modifies the resonance frequency of the SRR of the line. Nevertheless, this frequency is not influenced by the SRR of the tag, and hence it can be called $f_{r,SRR}$ (to point out that this frequency is only given by the SRR of the line). For $f_{r,SRR}$, the excursion (dynamic range) experienced by the transmission coefficient is close to 45 dB, i.e., -50 dB for the REF position, and roughly -5 dB for displacements above 1 mm. Finally, an intermediate frequency, called $f_{r,int}$, corresponding to the reflection zero frequency for a displacement of 0.4 mm, is also considered. In this case, the transmission coefficient as a function of the displacement exhibits a deep notch for the REF position, a maximum for a displacement of 0.4 mm, as expected, and then the transmission coefficient saturates to -15 dB as the relative displacement increases.

In view of Fig. 2.26, it is convenient to set the frequency of the feeding signal, or carrier frequency, to $f_c = f_{r,SRR}$. The reason is that for this frequency the maximum dynamic range is obtained. Nevertheless, by choosing the carrier frequency between $f_{r,BC\text{-}SRR}$ and $f_{r,SRR}$, the variation of the transmission coefficient is significant, and the value of the maximum transmission coefficient is close to the ideal value of 0 dB. Note that by setting f_c to values above $f_{r,SRR}$ the dynamic range is also considerable, but the value of maximum transmission progressively decreases as f_c increases, a situation that must be avoided to prevent deterioration of the modulation index. Thus, according to this analysis, the frequency of the feeding signal (carrier frequency) must satisfy $f_{r,BC\text{-}SRR} < f_c < f_{r,SRR}$, but, preferably, it must be as close as possible to $f_{r,SRR}$. This window for f_c is interesting from a practical viewpoint, since it is difficult to exactly predict the position of the optimum frequency, $f_{r,SRR}$. This analysis reveals that the proposed system is not very sensitive to frequency drifts, this being a relevant aspect.

One important parameter notably influencing the behavior/performance of the proposed chipless RFID system is the air gap distance. As the gap increases, the SRR-loaded line is more insensitive to the presence of the tag, since the coupling between the SRR of the line and the SRR of the tag decreases. By increasing the air gap, $f_{r,BC\text{-}SRR}$ increases, whereas $f_{r,SRR}$ remains constant. Figure 2.27 depicts the variation of $f_{r,BC\text{-}SRR}$ with the air gap. It can be seen that the window for f_c decreases as the air gap increases. Nevertheless, the window is significant up to reasonable gap distances. The reason of considering an air gap of 265 μm is related to the fact that with this separation between the tag and the reader, the frequency of maximum transmission

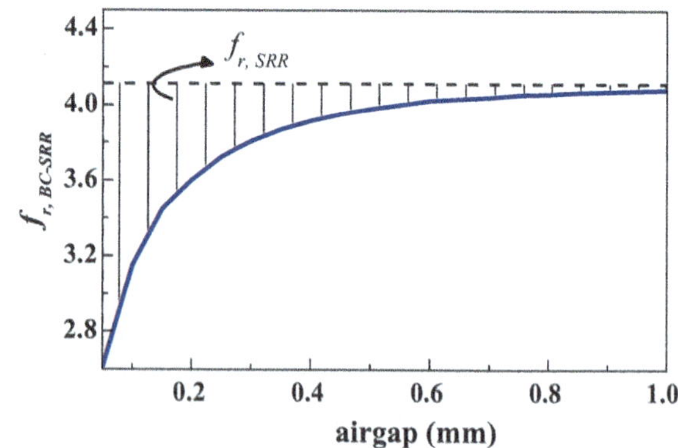

Fig. 2.27 Variation of $f_{r,BC\text{-}SRR}$ with the air gap separation. These results have been inferred by electromagnetic simulation. Reprinted with permission from [10]; copyright 2017 IEEE

of the SRR-loaded line without tag, $f_{r,SRR}$, coincides with the transmission zero frequency of the SRR-loaded line with perfectly aligned SRR on top of it. Thus, by tuning the frequency of the feeding signal to that frequency ($f_c = f_{r,SRR}$), the dynamic range is optimized.

2.3.2.2 Validation

System validation has been carried out by etching the tags in the perimeter of a circularly shaped substrate (with the aforementioned parameters), where four different 40-bit tags (each one implemented in a different quadrant of the circle) have been considered [10]. In these tags, the '0' state is obtained by the absence of resonant element. The photograph of these tags, as well as the photograph of the SRR-loaded line of the reader, is depicted in Fig. 2.28. Note that the 50 Ω access lines are bent in order to avoid mechanical friction between the port connectors and the tag during tag motion.

Tag reading has been carried out by attaching the tag to the rotor of the set-up depicted in Fig. 2.14 (see Sect. 2.3.1.2 for further details on such experimental set-up). The normalized envelope functions corresponding to the four considered codes

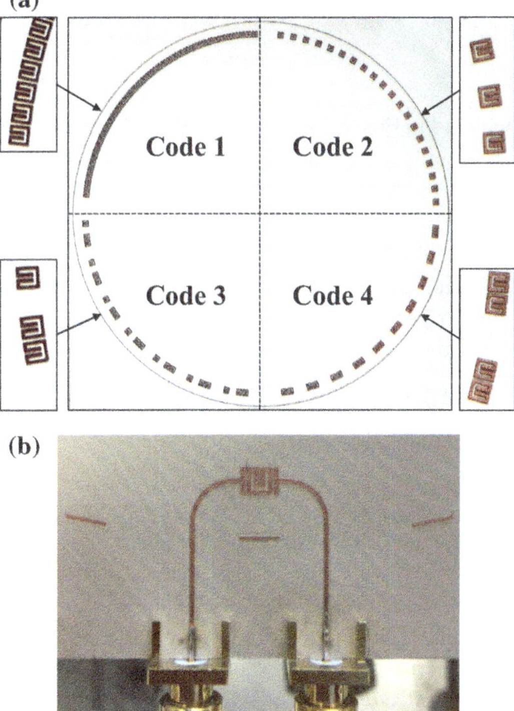

Fig. 2.28 Photograph of the fabricated **a** 40-bit chipless RFID tags. **b** Sensitive part of the reader

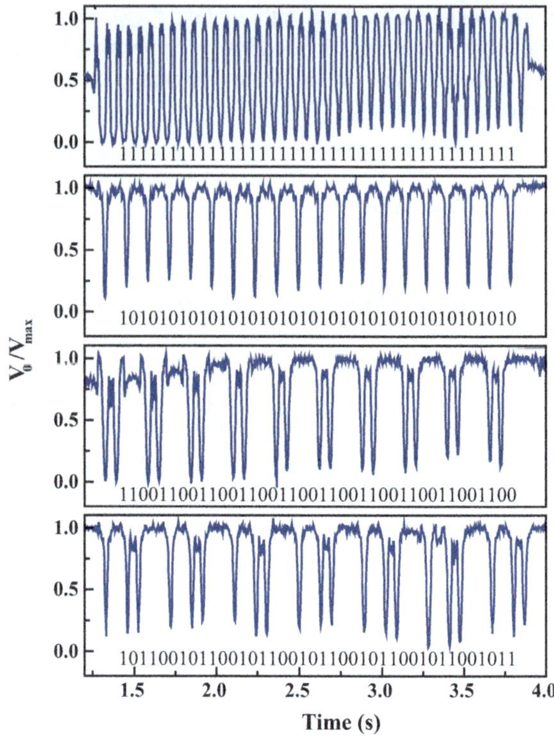

Fig. 2.29 Measured normalized envelope for the 40-bit fabricated tags. The depth of the dips, corresponding to the logic state '1', is significantly larger as compared to the system of Sect. 2.3.1.2. The angular velocity of the step motor is to 5.33 rpm. Reprinted with permission from [10]; copyright 2017 IEEE

are depicted in Fig. 2.29 (where the ID codes are indicated). Such envelope functions have been inferred by tuning the feeding interrogation signal to $f_c = f_{r,SRR} = 4$ GHz. It can be seen that the notches in the time response perfectly correlate with the logic state '1'. The significant dip depth indicates that the reader is very sensitive for the detection of the logic state '1', a necessary condition to implement a robust system for tag identification. The area of each 40-bit tag is as small as 4.75 cm^2, corresponding to a density per surface of DPS $= 8.4$ bit/cm^2. The per unit length density is DPL $=$ 2.97 bit/cm.

2.3.2.3   Tolerance Analysis

Let us now analyse the tolerance of the system against air gap variation and lateral misalignments. The excursion experienced by the transmission coefficient at $f_c = f_{r,SRR} = 4$ GHz with the air gap, when the SRRs of the line and tag are perfectly aligned, is depicted in Fig. 2.30a. The optimum gap spacing, providing the maximum variation of the transmission coefficient, is 265 μm. As anticipated before, the particularity of this gap separation is that, for perfectly aligned SRRs, the transmission zero frequency exactly coincides with the reflection zero frequency of the structure without tag on top of it, i.e., $f_{r,SRR}$, which is in turn the carrier frequency. It

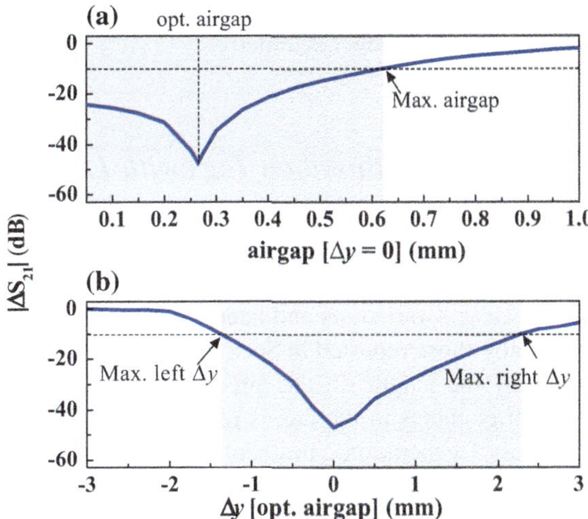

Fig. 2.30 Tolerance analysis for vertical and lateral displacement of the tag on the maximum variation of the transmission coefficient. **a** Effects of air gap variation. **b** Effects of lateral shift. The mechanical system used to provide relative motion between tag and reader must guarantee lateral and vertical misalignments within the indicated tolerance limits. These results have been inferred by electromagnetic simulation. Reprinted with permission from [10]; copyright 2017 IEEE

should be highlighted, however, that this optimum gap space has been calculated for a particular carrier frequency, i.e., $f_c = f_{r,SRR}$. If the carrier frequency is slightly below $f_{r,SRR}$, then it is necessary to decrease the gap in order to allocate the transmission zero, for the perfectly aligned SRRs, at the position of the carrier frequency. Thus, the optimum gap separation depends on the carrier frequency, but such optimum gap separation increases with f_c. Since f_c should not be chosen to be higher than $f_{r,SRR}$, it follows that $f_{r,SRR}$ is the optimum carrier frequency, provided the optimum gap distance is the largest one within the interval. Nevertheless, small variations of f_c in the vicinity of $f_{r,SRR}$ (with the gap set to the optimum value at $f_{r,SRR}$) do also give significant excursions in the transmission coefficient. In view of Fig. 2.30a, with the interrogation signal tuned to $f_c = f_{r,SRR} = 4$ GHz, up to 0.62 mm air gap separation can be tolerated, provided a minimum excursion of the transmission coefficient of 10 dB is required. This tolerance is substantially high, as compared to the one of the system described in Sect. 2.3.1.

In order to analyze the tolerance against lateral shifts of the tag with regard to the line axis, Fig. 2.30b depicts the maximum variation of the transmission coefficient for the optimum frequency $f_{r,SRR}$ and the optimum gap separation at this frequency (265 μm). As can be seen, laterally shifting the tag degrades the excursion of the transmission coefficient. However, by considering a tolerance limit of -10 dB for a reliable reading operation, it follows that lateral shifts between -1.3 and $+2.3$ mm are within the allowable limits for misalignment in the transverse direction. These values are very reasonable on account of the considered SRR dimensions. It is worth mentioning that the tolerance interval is not symmetric. The reason is the lack of symmetry of the structure with regard to the line axis. However, due to the symmetry with regard to the midplane between the input and the output port for perfectly aligned SRRs, it follows that tag displacement in the positive or negative direction, from the

REF position, along the line axis is undistinguishable. For this reason, the curves of Fig. 2.26 exhibit perfect symmetry.

2.3.3 Systems Based on Tags with Linearly-Shaped Strips

The previous time-domain signature chipless-RFID systems, based on tags implemented with square (or rectangular) shaped resonant elements, are quite tolerant against air gap variations and lateral misalignments between the tag and the reader (especially those reported in Sect. 2.3.2). However, the DPL in such systems, determined by the length (in the direction of the chain axis) and separation between resonant elements in the tag, is relatively small (e.g., to achieve the 96 bits of the EPC class 1 generation 2 protocol of UHF-RFID, a 32.3-cm tag is needed).

In this section, we report a different time-domain signature chipless-RFID approach, where the tags consist of a chain of linearly-shaped narrow and tiny separated strips transversally oriented to the chain axis. By this means, the chain period can be significantly reduced, as compared to the realizations of Sects. 2.3.1 and 2.3.2, and hence the DPL of the tags can be substantially improved, as it will be demonstrated. Indeed, the optimum DPL would be the one achievable by assigning the minimum slot (s_m) and strip (w_m) width of the available fabrication technology to the separation (s) and width (w), respectively, of adjacent strips in the tag. Typical (conservative) values in PCB and printing technologies are $s_m = w_m = 0.2$ mm. Therefore, by fabricating the tags with $s = s_m$ and $w = w_m$, the resulting optimum data density per length is found to be $\text{DPL}_{opt} = 25$ bit/cm.

For tag reading, a near-field system able to distinguish two types of strips, those associated to the '0' state and those associated to the '1' state, or, alternatively, the presence or absence of strip at the predefined tag positions, is needed. Nevertheless, since the tags are based on linear strips, the use of linear resonant elements to read the tag ID codes seems to be a good option. Therefore, readers based on both linear quarter-wavelength and half-wavelength resonators, are considered in this section. As it will be later shown, tags with strip separation and width of $s = 0.2$ mm and $w = 0.4$ mm, respectively, corresponding to a linear density of $\text{DPL} = 16.7$ bit/cm, are correctly read with both readers. In the last part of the chapter, a reader based on a double-stub loaded line is also reported. With this reader, very short strips can be detected. Thus, tags with good DPL and DPS are readable.

2.3.3.1 Reader Based on a Half-Wavelength Resonator

Let us first study the system based on a reader implemented by a half-wavelength resonator. Particularly, an order-1 capacitively coupled resonator microstrip bandpass filter is considered as the sensitive (active) part of the reader (Fig. 2.31) [13]. Tag reading with this system is based on the variation experienced by the transmission coefficient of this structure when a linear metallic strip (one of the strips of the

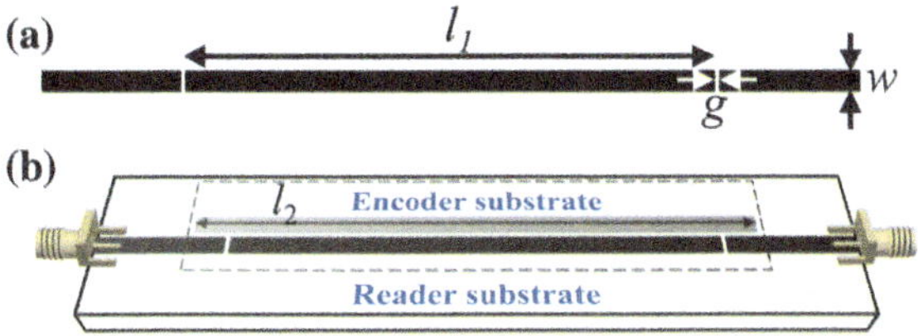

Fig. 2.31 Layout of the half-wavelength resonator based reader (**a**), and three-dimensional view of the reader with a strip on top of it and perfectly aligned (**b**). Dimensions are (in mm): $l_1 = 21.2$, $l_2 = 22.4$, $g = 0.2$ and $w = 0.4$. The considered reader substrate is the *Rogers RO4003C* with dielectric constant $\varepsilon_r = 3.55$, thickness $h = 0.81$ mm, and loss tangent $\tan\delta = 0.0021$. The considered tag substrate is the *Rogers RO4003C* with dielectric constant $\varepsilon_r = 3.55$, thickness $h = 0.204$ mm, and loss tangent $\tan\delta = 0.0021$. Reprinted with permission from [13]; copyright 2019 IEEE

tag) is located on top of the resonant element of the reader at short distance (see Fig. 2.31b). The transmission coefficient of the bare structure, i.e., without strip on top of it (but with the presence of the tag substrate), exhibits a resonance peak at roughly the frequency where the length of the line is equivalent to a half wavelength. Actually such frequency depends also on the pair of gaps present between the central line section and the input and output access lines, since such gaps provide a non-negligible reflection phase due to its capacitive effect. The resonance condition can be written as

$$\varphi_1 + \varphi_2 + 2\beta l_1 = 2\pi n \tag{2.4}$$

where φ_1 and φ_2 are the reflection phases in gaps 1 and 2, respectively, βl_1 is the electrical length of the central transmission line section, β and l_1 being the phase constant and the physical length of such section, and $n = 1, 2, 3, \ldots$. From (2.4), it follows that for an uncoupled resonator (with $\varphi_1 = \varphi_2 = 0$), the first resonance ($n = 1$) occurs when $\beta l_1 = \pi$ (or $l_1 = \lambda/2$, where λ is the guided wavelength), as it is well known.

From (2.4), it is clear that the resonance condition can be altered either by modifying the reflection phases or by perturbing the phase constant (or both simultaneously). This can be achieved by the presence of a linear (straight) strip on top of the half-wavelength resonator, parallel oriented to the line axis, with a length extending over the gap positions, as indicated in Fig. 2.31b. Figure 2.32 depicts the response of the bare reader inferred from full wave electromagnetic simulation (using *Keysight Momentum*), as well as the response of the reader with a perfectly aligned strip on top of it, where it can be appreciated that the resonance peak is shifted to lower frequencies. Actually, the responses of the bare and loaded reader have been obtained by considering the presence of adjacent strips separated 0.2 mm (corresponding to a tag period of 0.6 mm, as anticipated before). The reason is that these are the worst conditions regarding the excursion of the transmission coefficient at the interrogation signal frequency, to be determined.

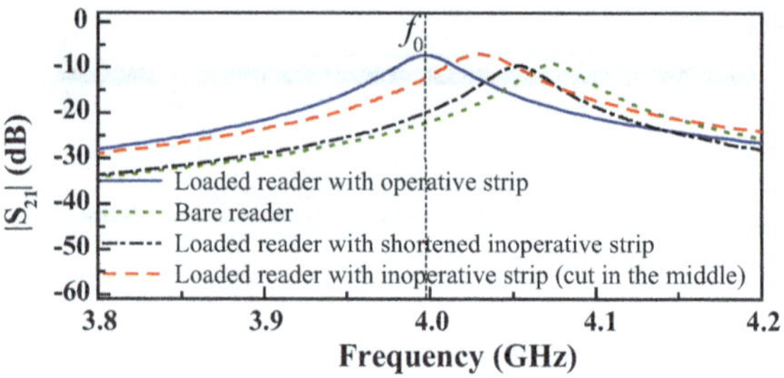

Fig. 2.32 Simulated frequency responses (including losses) of the bare reader, and reader loaded with perfectly aligned strips under different conditions, including operative strip with nominal length, inoperative strip with cut in the middle position, and inoperative strip with shortened length. The considered air gap, or vertical distance between the tag and the reader is 0.2 mm. In all the simulations, a pair of adjacent operative strips has been considered. Reprinted with permission from [13]; copyright 2019 IEEE

Obviously, the binary states can be associated to the presence or absence of strip in the predefined position in the tag. Alternatively, the '0' logic state can be achieved by making the corresponding strips inoperative, i.e., unable to substantially modify the response of the bare reader. For that purpose, two options can be considered. The first one is to cut the strip in the central position (see Fig. 2.33c). The transmission coefficient, also depicted in Fig. 2.32, is different to the one corresponding to the presence of a functional strip on top of the reader, but it is far from the response of the bare reader. The reason is thought to be due to the fact that the main cause of resonance variation is the modification of the effective capacitance of the gap, which is not altered by implementing a cut in the central position of the strip. The second approach consists of shortening the strips by the extremes (see Fig. 2.33d), providing a strip length identical to the one of the resonant element of the reader. By this means, the effect of the strip on the gap capacitance is expected to be smaller, and, consequently, a response closer to the one of the bare reader is expected. This is confirmed in Fig. 2.32, where such response is also depicted. According to these results, writing the '0' logic state in the tags should be done either by eliminating the corresponding

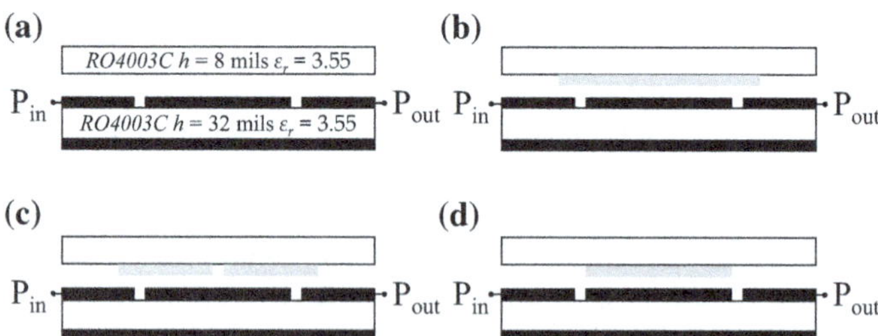

Fig. 2.33 Cross-sectional view of the strips and reader corresponding to the four responses depicted in Fig. 2.31. **a** Bare reader. **b** Loaded reader with operative strip. **c** Loaded reader with inoperative strip with a cut in the central position. **d** Loaded reader with shortened inoperative strip

strip or by shortening it. Note that, in both cases, the resulting response exhibits a significant excursion of the transmission coefficient at the resonance frequency of the reader loaded with operative strip, f_0, indicated in Fig. 2.32.

By tuning the frequency of the interrogation signal to $f_c = f_0$, tag reading with a system similar to those used in the systems reported in the previous sections can be carried out. Note, however, that, contrary to the previous systems, the '1' logic states provide maximum signal transmission (maximum in the envelope function), whereas signal transmission is minimized when a strip is not present on top of the reader or when it is present but shortened ('0' state). Figure 2.34 depicts the transmission coefficient at $f_0 = 4$ GHz, when two 5-bit tags with the indicated codes are displaced over the reader (the '0's are achieved by shortening the corresponding strips). The considered air gap is 0.2 mm.

The validation of this system is carried out by reading 100-bit tags with different ID codes. The sensitive part of the reader and the 100-bit tag with all bits set to '1' (long strips) are depicted in Fig. 2.35. At this point it is worth to mention that the minimum strip width of the tags able to provide the necessary excursion in the transmission coefficient to distinguish the binary states is 0.4 mm, and this is the reason why the chain period has not been further reduced. That is, with the period of the designed tags (0.6 mm), the DPL is not the optimum one (see above), but the resulting value (DPL = 16.7 bit/cm) is very good, and provides a tag length of only 6 cm. The measured normalized envelope functions for the different considered codes, inferred from the original fabricated 100-bit tag by shortening certain strips, are depicted in Fig. 2.36.

An important aspect related to system robustness, is the readability of the system when it is subjected to mechanical vibrations or misalignments. Thus, the study of the effects of misalignment/orientation and vertical distance (gap) variation between the reader and the strip chain is convenient. A case example consisting of a 5-strip chain with alternating operative and inoperative strips (identical to the one of Fig. 2.34) is considered. Figure 2.37a depicts the simulated responses obtained by displacing the

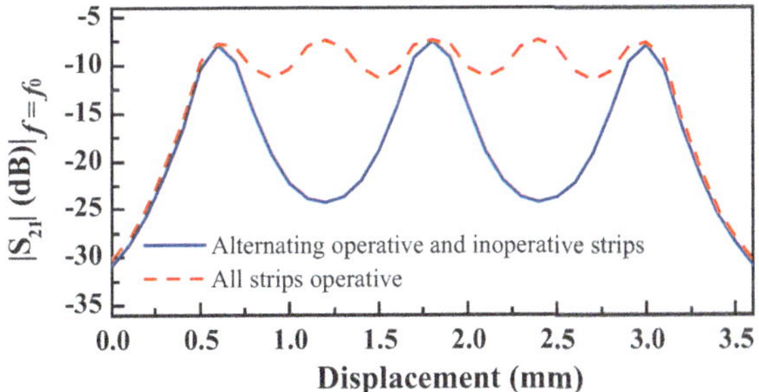

Fig. 2.34 Transmission coefficient at f_0 as a function of the relative displacement between the reader and the strip chain. Two cases are considered, i.e., with all strips operative, and with operative and inoperative (shortened) strips alternating. Reprinted with permission from [13]; copyright 2019 IEEE

Fig. 2.35 Photograph of the strip chain (**a**) and sensitive part of the reader (**b**). The space occupied by the tag is 60 mm × 22.4 mm

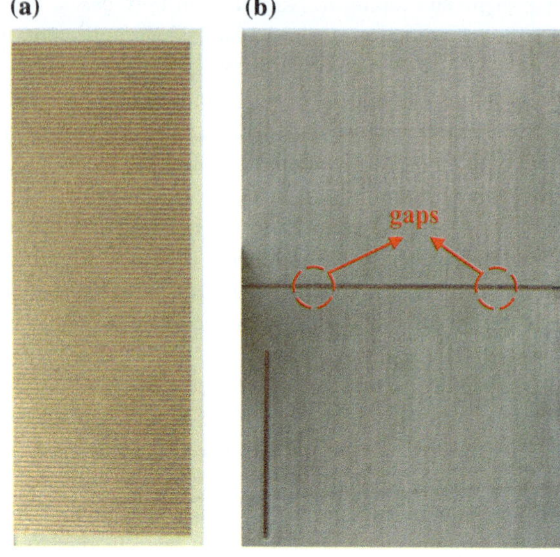

Fig. 2.36 Measured normalized envelope function corresponding to the tags with the indicated codes (shown in the insets, where zoom views are included for clarity). Note that the minimum bit sequence that repeats is '10', '1110' and '101000' from top to bottom. Reprinted with permission from [13]; copyright 2019 IEEE

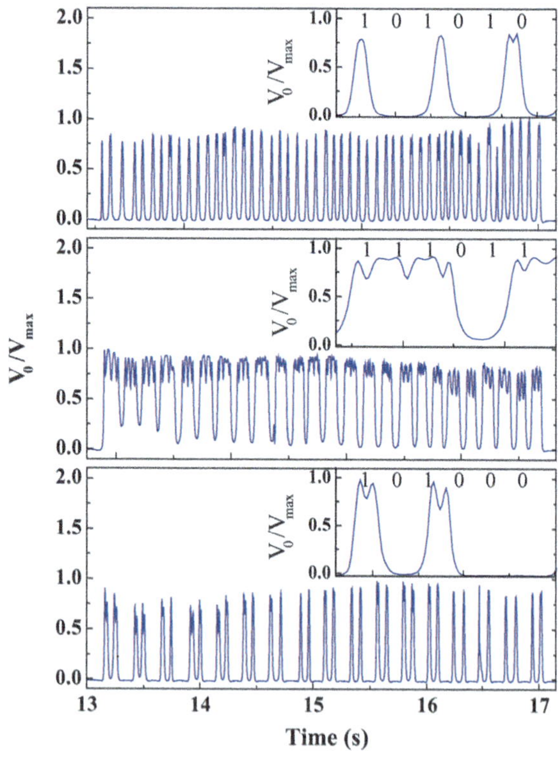

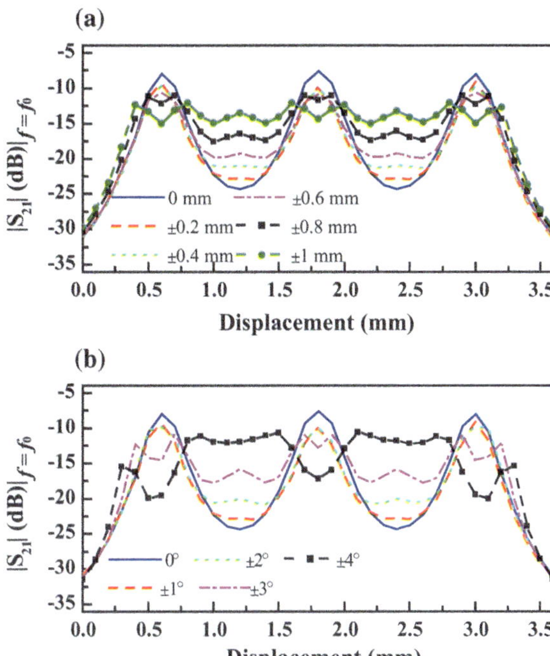

Fig. 2.37 Effects of lateral misalignment (**a**) and lack of alignment by rotation (**b**) on the transmission coefficient at f_0 when the strip chain is displaced over the reader. The considered air gap is 0.2 mm. Reprinted with permission from [13]; copyright 2019 IEEE

strip chain over the reader for different lateral misalignments. The system is tolerant up to lateral displacements of roughly ±0.8 mm (above that value, the operative and inoperative strips cannot be distinguished). Nevertheless, it should be mentioned that by elongating the strips (with the penalty of larger size), the above tolerance in lateral displacement can be improved.

Let us now analyze the effects of misalignment in relative orientation (rotation) between the strip chain and the reader.[11] For that purpose, the response inferred by displacing the strip chain over the reader by considering different angles between the line axis (reader) and the axis of the strips in the chain (the perfect situation corresponding to an angle of 0°), is reported in Fig. 2.37b. The system is tolerant up to an angle of roughly ±3°, which is reasonable from a practical viewpoint. Better tolerance against lack of alignment by rotation can be obtained, but at the expense of wider line and strips (which represents a penalty in terms of data density for chipless-RFID systems and resolution for linear displacement measurements).

Figure 2.38 shows the effects of the air gap on the response of the system. The structure has been optimized for a nominal air gap of 0.2 mm, where the maximum excursion of the transmission coefficient results. Increasing the air gap (Fig. 2.38a) reduces such excursion, but reasonable values are obtained within the considered margin (up to 0.23 mm). Figure 2.38b depicts the responses inferred by decreasing

---

[11] Lack of alignment by a relative rotation between the tag and the reader may be specially critical in tags based on very narrow strips.

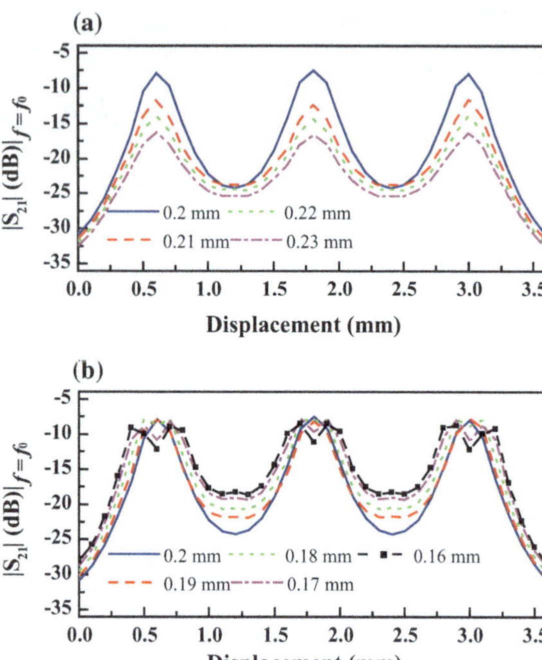

Fig. 2.38 Effects of air gap on the transmission coefficient at f_0 when the strip chain is displaced over the reader. **a** Air gaps above 0.2 mm. **b** Air gaps below 0.2 mm. Reprinted with permission from [13]; copyright 2019 IEEE

the air gap. Even for a vertical distance between the strip chain and the reader as small as 0.16 mm, the excursion experienced by the transmission coefficient is reasonable. Consequently, system tolerance against vertical distance is at least in the range 0.16–0.23 mm.

It is worth mentioning that for air gaps of 0.16 mm and 0.17 mm, a double peak in the response is visible. This is due to the fact that by decreasing the air gap, the maximum in the frequency response of the structure (coincident with f_0 in Fig. 2.32, where the air gap is 0.2 mm) is displaced to the left. Under these conditions, two separated maxima are expected when the strip chain is displaced above the reader and the frequency of the injected signal is f_0. Obviously, such maxima are visible (double peak) when the frequency of the maximum transmission coefficient lies significantly below f_0, and this occurs for sufficiently small air gaps (such as 0.16 and 0.17 mm in the case study).

2.3.3.2 Reader Based on a Quarter-Wavelength Resonator

A quarter wavelength resonator can be implemented by means of an open-ended stub. A transmission line loaded with said stub, exhibits transmission zeros at the fundamental resonance of the stub (the one where the length of the stub is a quarter wavelength) and at the odd harmonics. In the vicinity of these frequencies, the stub can be modeled by a grounded shunt connected series resonator. Figure 2.39a depicts

Fig. 2.39 Transmission line loaded with a shunt resonator (**a**), the same structure with a resonant tank coupled to it (**b**), and response of both structures inferred from *Keysight ADS* (**c**). Element values are $L = 1.9$ nH, $C = 1.1$ pF, $L_R = 1.9$ nH, $C_R = 1.1$ pF and $M = 0.6$ nH. Reprinted with permission from [14]; copyright 2018 IEEE

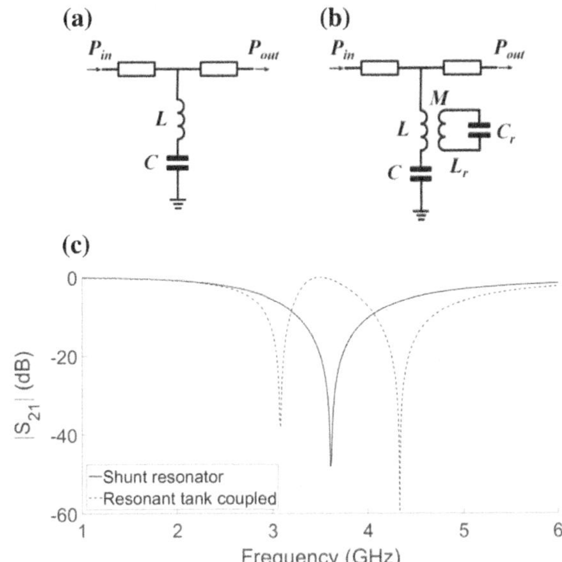

the model of the stub-loaded line in the vicinity of the fundamental resonance. Let us now consider that a resonant element is coupled to the shunt resonator, as depicted in Fig. 2.39b, where magnetic coupling through the mutual inductance M is considered. The effect is a split in the original transmission zero frequency, as depicted in Fig. 2.39c.[12] It can be seen that the variation experienced by the transmission coefficient in the vicinity of the resonance frequency of the unloaded resonator, f_0, is significant. Consequently, this splitting effect can be exploited for coding purposes. Particularly, linear half-wavelength resonators can be used in the tag side, as long as such resonant elements can be excited by means of a quarter-wavelength shunt stub. Hence, the considered reader in this implementation is a stub loaded line.

The layout of the sensitive part of the reader and a three-dimensional view of the reader and the tag are depicted in Fig. 2.40. The first chipless-RFID system based on this approach was reported in [14], by tuning the dimensions of the half-wavelength tag resonators to the first resonance frequency of the stub-loaded line of the reader. This system was validated in [14], but the effects of adjacent resonators on the transmission coefficient were found to be significant by reducing the inter-resonator distance, s, to small values (e.g., 0.2 mm). To alleviate this limitation, the distance between adjacent tag strips was set to $s = 1.6$ mm in [14], corresponding to a chain period of 2 mm (provided the strip width was set to $w = 0.4$ mm). The resulting data density per length (DPL $= 5$ bit/cm) is far from the value achieved in the system presented in sub-Sect. 2.3.3.1.

In [15], it was demonstrated that by tuning the half-wavelength resonator strips of the tag to the second resonance of the stub (where the length of the stub is 3/4λ,

---

[12]Frequency splitting occurs regardless of the type of coupling, i.e., magnetic, electric, or mixed.

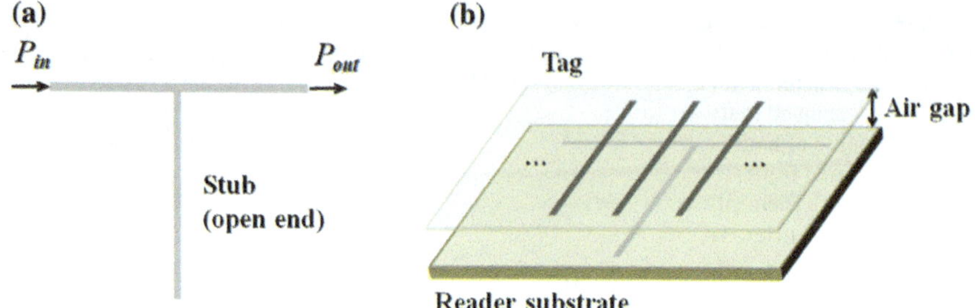

Fig. 2.40 Layout of the active part of the reader (**a**) and 3D view of the loaded reader with one of the half-wavelength resonators of the tag aligned with the reader stub (**b**). The length and width of the stub are 26 mm and 0.2 mm, respectively. The width of the transmission line is 0.6 mm (50 Ω line). The separation between adjacent resonators in the tag chain is 0.2 mm and resonator dimensions are 34 mm $\times$ 0.4 mm (so that the tag period is 0.6 mm). The tag chain has been implemented on the *Rogers RO4003C* substrate with thickness $h = 0.2$ mm and dielectric constant $\varepsilon_r = 3.55$. The stub-loaded line has been implemented on the *Rogers RO3010* substrate with thickness $h = 0.635$ mm and dielectric constant $\varepsilon_r = 10.2$

λ being the wavelength), the proximity effects of the tag strips are less severe, and the system was validated by considering a tag period of 0.6 mm (i.e., with DPL = 16.7 bit/cm), identical to the period of the tags reported in the previous sub-section. By considering the dimensions and substrate parameters indicated in the caption of Fig. 2.40, the frequency response of the bare reader and the response of the reader loaded with a single half wavelength resonator tuned to the second resonance of the stub, are found to be those depicted in Fig. 2.41 (with an air gap of 0.3 mm). Frequency splitting occurs at the second resonance of the stub, but not at the first resonance, where the length of the tag strips is one third of half a wavelength (or $\lambda/6$). Note, however, that by tuning the half wavelength resonators to exhibit the first

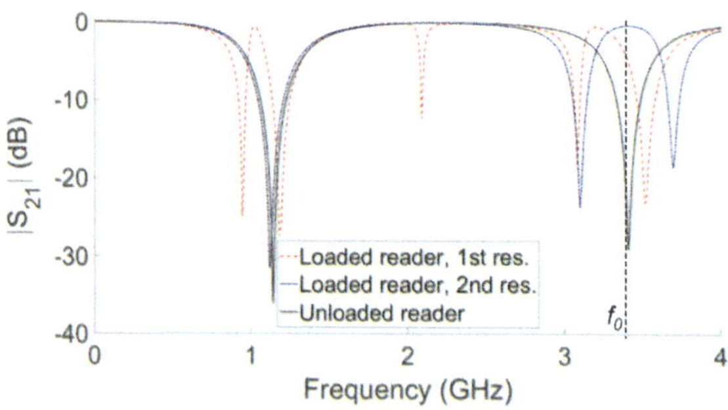

Fig. 2.41 Magnitude of the transmission coefficient (S_{21}) of the unloaded reader and reader loaded with a single half-wavelength resonator perfectly aligned. For the loaded reader, two different cases have been considered in regard to the half-wavelength resonator, i.e., either tuned at the first (fundamental) or second resonance of the stub. The considered air gap is 0.3 mm. These results have been inferred by electromagnetic simulation, using *Keysight Momentum*. Reprinted with permission from [15]; copyright 2019 IEEE

resonance at the first resonance of the stub (conveniently modifying the strip length), splitting occurs at both frequencies, as can be also seen in Fig. 2.41.

In view of Fig. 2.41, the interrogation signal must be set to $f_c = f_0 = 3.47$ GHz. Figure 2.42 depicts the variation of the transmission coefficient at this frequency for two different 7-bit tags (with the indicated codes), by considering different airs gaps. From this figure, it can be concluded that the tags can be correctly read, as far as there is a significant difference in the transmission coefficient between the binary states. Moreover, the system is quite tolerant against air gap variations, at least within the considered limits. Concerning lateral misalignments between the tag and the reader, Fig. 2.43 depicts the transmission coefficient for the code '1101001' parameterized by the relative lateral shift. Note that, due to the lack of axial symmetry of the reader, the responses corresponding to positive and negative displacement of the same magnitude are not identical (it has been considered that positive displacement corresponds to lateral displacement of the tag in the direction of the stub). In view of this figure, it can be considered that the system is quite tolerant against lateral displacements.

Figure 2.44 shows the photograph of the reader and the photograph of a fabricated 100-bit tag (with all bits set to '1'), whereas the corresponding normalized envelope function can be seen in Fig. 2.45. From this tag, several codes have been generated by cutting certain half-wavelength resonators by the central position (in the simulations depicted in Fig. 2.42, the '0' states were also achieved by eliminating a small portion of the strip, equivalent to cut it). The time-domain responses of these tags are also

Fig. 2.42 Variation of the transmission coefficient at the interrogation signal frequency that results from a relative displacement between the tag and the reader for the indicated codes. Several air gaps are considered. Reprinted with permission from [15]; copyright 2019 IEEE

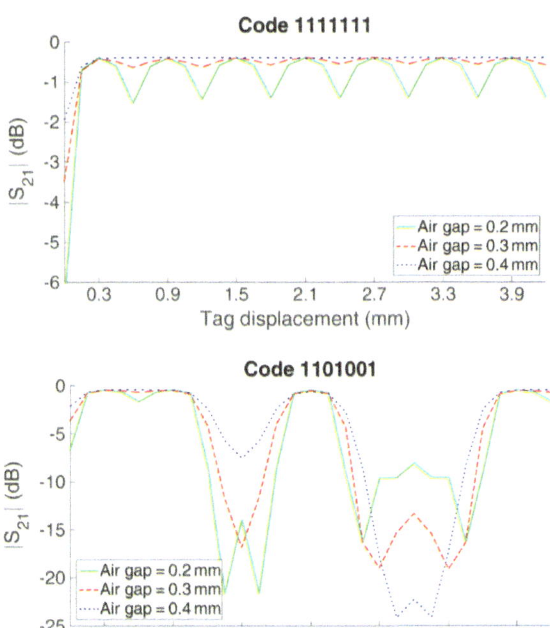

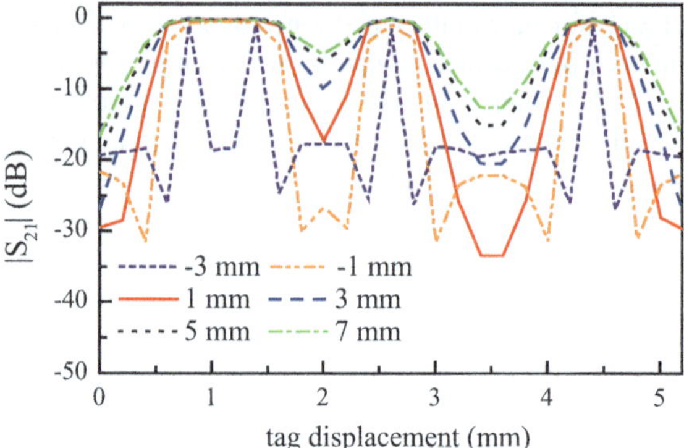

Fig. 2.43 Variation of the transmission coefficient at the interrogation signal frequency that results by displacing the tag over the reader at 0.3 mm (air gap), corresponding to the code '1101001'. Several lateral displacements (positive and negative) are considered

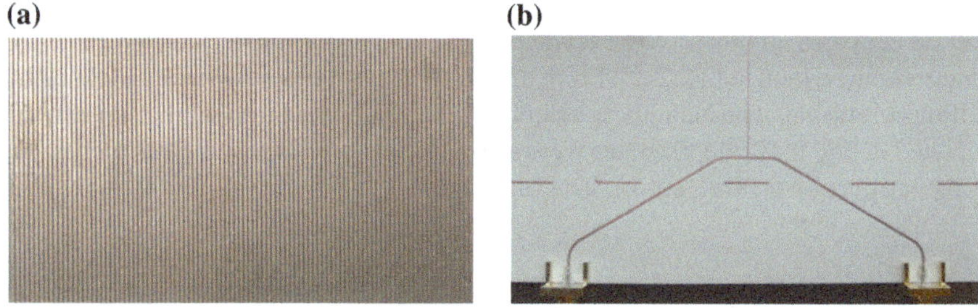

Fig. 2.44 Photograph of the fabricated 100-bit tag with all bits set to the logic state '1' (a) and reader (b). The space occupied by the tag is 60 mm × 34 mm

depicted in Fig. 2.45, and reveal that dips corresponding to the '0' states are present. Thus, these results validate this time-domain signature chipless-RFID system based on tags and reader implemented with linear strips and a stub-loaded line, respectively.

2.3.3.3 Reader Based on a Double-Stub Loaded Line

The tags based on linear strips reported in the previous subsections exhibit a reasonable shape factor, with very good DPL. However, the DPS in such tags is not as good as desired due to the long length of the linear strips (this aspect was already discussed in Sect. 2.2). Such strip lengths are necessary in order to correctly read the tags with the considered readers (based on linear resonators). In this subsection, a different strategy that allows for the implementation of tags with much shorter strips is reported [32]. The resulting tags exhibit, therefore, not only a good DPL, but also a competitive DPS.

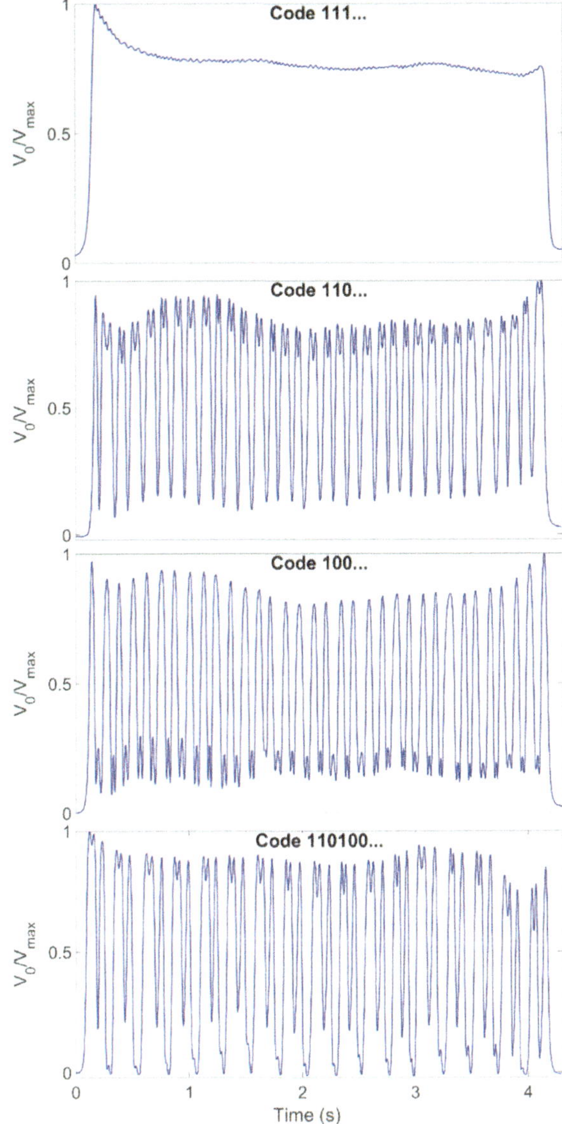

Fig. 2.45 Measured normalized envelope for the 100-bit tags with the indicated codes. For each code, the repeating period is shown. The reason of the variation of the maximum voltage in the figures is related to the fact that absolute uniformity in the gap space between the tag and the reader cannot be guaranteed. Reprinted with permission from [15]; copyright 2019 IEEE

The considered reader is based on a microstrip line loaded with two open-ended folded stubs separated a tiny distance by their extremes (the layout and relevant dimensions are given in Fig. 2.46). With such reader, it is possible to detect the presence of relatively short strips (at least as compared to those of the previous tags) when they are located on top of the sensitive part of the reader (the extremes of the stubs). The layout of the tag (with all-functional elements) compatible with this reader system is also depicted in Fig. 2.46. The strip length is as short as 6.4 mm,

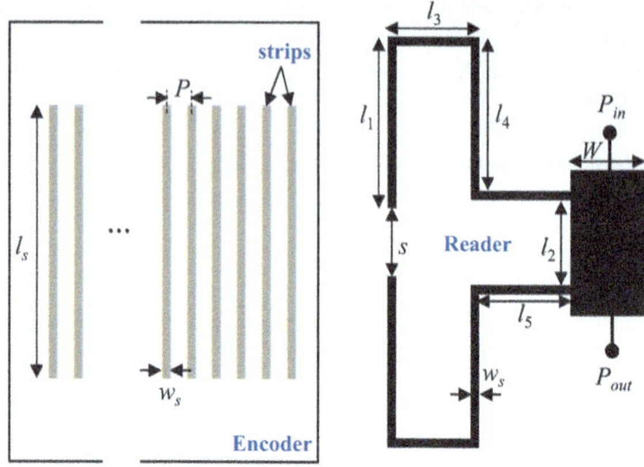

Fig. 2.46 Topology of the double-stub reader and tag, and relevant dimensions (in mm). $W = 1.81$, $l_0 = l_2 + w_s = 2.2$, $l_1 = 3.8$, $l_3 = 3.3$, $l_4 = 2.1$, $l_5 = 2.1$, $s = 1.60$, $P = 0.60$, $w_s = 0.20$ and $l_S = 6.4$. Reprinted with permission from [32]; copyright 2019 IEEE

whereas the strip width and separation is 0.2 and 0.6 mm, respectively. With these tag dimensions the resulting data density is DPL = 12.5 bit/cm and DPS = 26.04 bit/cm^2. The DPL is somehow inferior to the one of the tags based on linear strips analyzed before, but the DPS is by far larger than the one of any of the time-domain signature barcodes considered in this chapter.

System functionality is based on the enhancement of the capacitive coupling between the stubs when a functional strip of the tag is located on top of the sensitive region at short distance (air gap). For the bare reader (without strip on top of it), the coupling between the stubs is very small. However, the presence of a relative short strip on top of the reader enhances substantially the inter-stub coupling. Indeed, considering identical stubs, if inter-stub coupling is not present, a single transmission zero at the frequency where the stub length is a quarter-wavelength, and at the odd harmonics, is expected. However, stub coupling by the presence of a functional strip on top of the reader produces a frequency splitting in the transmission zeros. Moreover, the response of the reader exhibits a pole that depends on the distance between the contact points of the stubs to the host line (l_2). Such pole also shifts down when inter-stub coupling is enhanced. Thus, by appropriately designing the reader, it is possible to achieve a large excursion in the transmission coefficient for a certain frequency. The specific strategy is to force the pole of the bare reader to be identical to the first transmission zero of the loaded (with a strip) reader, and to set the interrogation signal, f_0, to that frequency. The specific design procedure is explained in detail in [32]. The photograph of the fabricated reader and tag with all functional resonators is depicted in Fig. 2.47, whereas the frequency response of the reader with and without strip on top of the sensitive region is shown in Fig. 2.48 (the convenient frequency position of the harmonic interrogation signal is indicated in the figure).

The validation of the system has been carried out by obtaining the envelope functions of three different 100-bit tags. In all the cases, a periodic ID code has been considered, particularly, one with all bits set to '1' logic state, i.e., '11111…', one with the sequence '101010…', and one with the sequence '110110110…'. After the

Fig. 2.47 Photograph of the fabricated reader (**a**) and 100-bit tag with all bits set to '1' logic state (**b**). The reader has been fabricated on the *Rogers RO4003C* with thickness $h = 0.81$ mm, dielectric constant $\varepsilon_r = 3.55$, and dissipation factor $\tan\delta = 0.0021$. For the tag, the considered substrate is the same as the reader but with thickness $h = 0.203$ mm. The space occupied by the tag is 60 mm × 6.4 mm

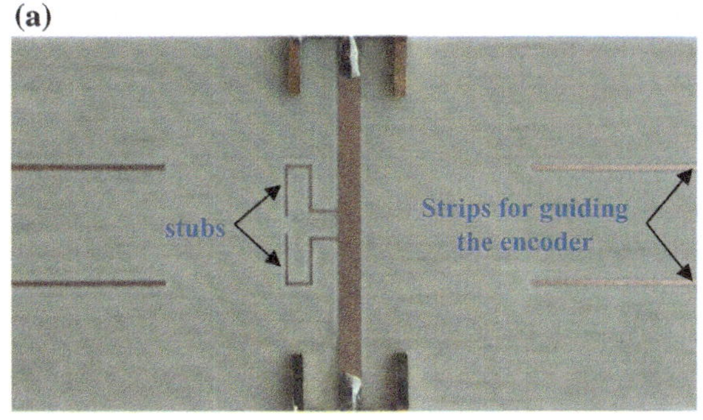

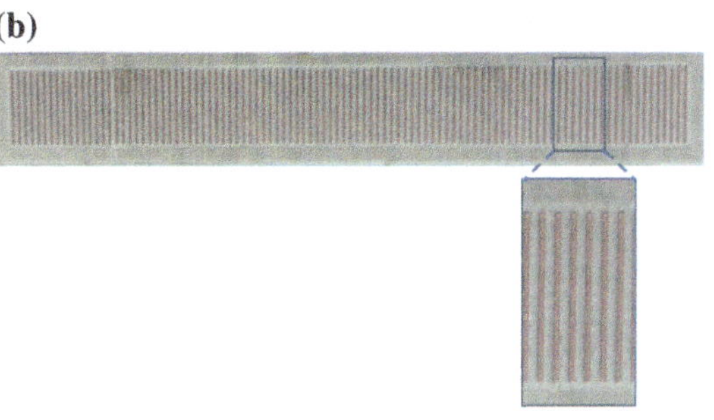

Fig. 2.48 Measured and simulated responses of the bare reader and reader loaded with a tag strip on top of it. The considered air gap is 0.2 mm. Reprinted with permission from [32]; copyright 2019 IEEE

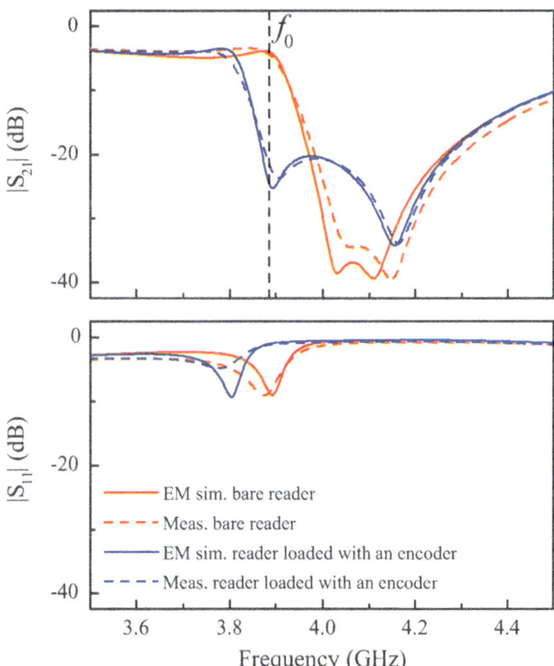

fabrication of the tag with all bits set to '1', the two additional tags have been obtained from this one by programming them, i.e., by cutting those strips corresponding to the logic state '0'. By this means, these strips are made inoperative, and roughly the same effect as the absence of strip on top of the sensitive part of the reader is achieved (tag programming is studied in more detail in the next chapter). The envelope functions of these encoders are depicted in Fig. 2.49, where it can be appreciated that the different ID codes are correctly read. The interrogation signal frequency has been set to $f_0 = 3.88$ GHz, the frequency of the pole for the bare reader. The dynamic margin, or separation between the voltage levels corresponding to the two states, is high. In view of Fig. 2.49 (where clearly separated voltage levels between the binary states can be appreciated), the functionality of this approach for the implementation of near-field chipless-RFID systems is validated.

Concerning the robustness of the system against air gap variations, and lateral or angular misalignment of the tag, a tolerance analysis is convenient. Such analysis is

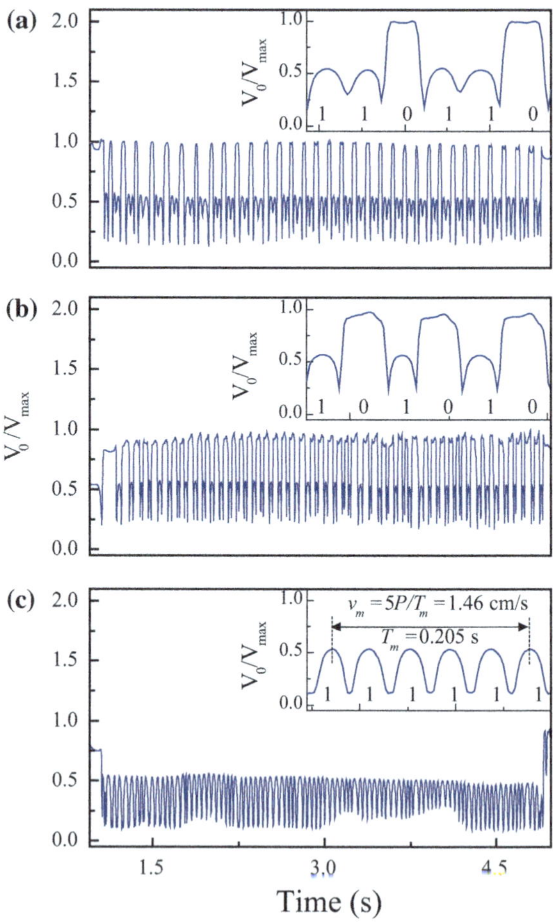

Fig. 2.49 Measured normalized envelope function corresponding to the indicated codes. Reprinted with permission from [32]; copyright 2019 IEEE

carried out at simulation level due to the difficulty to accurately control the involved geometrical parameters (namely, air gap separation and lateral or angular displacement of the tag). For that purpose, a 5-bit sequence with alternating binary states is considered. The effects of the air gap variation are depicted in Fig. 2.50a. The system has been optimized to operate with an air gap corresponding to the nominal value (0.2 mm), where the excursion experienced by the transmission coefficient between the two binary states is maximum. Slightly increasing or decreasing the air gap distance reduces such excursion. According to the sensitivity of the transmission coefficient with the air gap, the mechanical guiding system for encoder motion should guarantee small variations in the air gap separation. Further robustness against the effects of the air gap can be achieved, but at the expense of larger encoder dimensions. The effects of lateral misalignment of the encoder are shown in Fig. 2.50b. According to these results, lateral displacements up to ±1 mm provide excursions in the transmission coefficient better than 8 dB. With this variation, encoder reading from the AM signal generated at the output port of the reader in response to the interrogation signal seems feasible. Finally, Fig. 2.50c shows the effects of lack of alignment by tag rotation, where the responses up to an angle of ±6°, with step increments of ±1°, are depicted. According to the resulting transmission coefficient variations between the two binary states, the system is tolerant up to an angle of at least ±3° (providing an excursion of the transmission coefficient of roughly 9 dB or better).

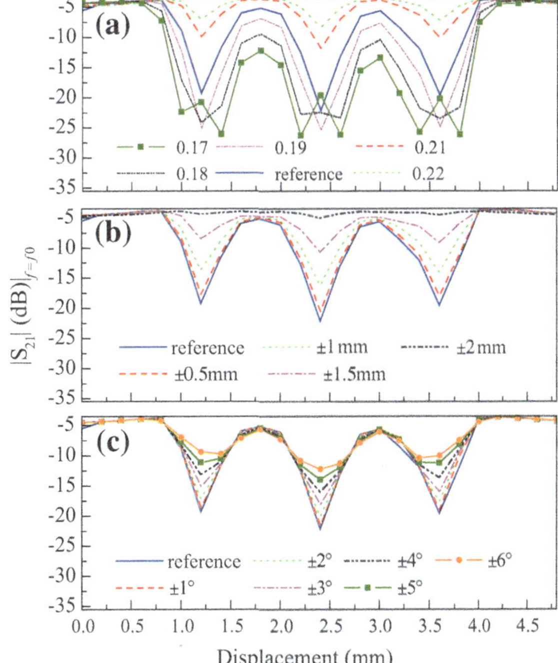

Fig. 2.50 Tolerance analysis. a Effects of the air gap variation. b Effects of lateral misalignment. c Effects of misalignment by encoder rotation. The curves correspond to the transmission coefficient at the frequency of the interrogation signal, as the 5-bit encoder with ID code '10101' is displaced over the reader. Reprinted with permission from [32]; copyright 2019 IEEE

The tolerances of this reader-tag system (based on a double-stub loaded line reader) against lateral and angular misalignments are reasonable from a practical viewpoint. That is, the angular and position accuracies of the encoder guiding system in the plane of the encoder do not need to be extremely good, according to the reported tolerances. Concerning the accuracy in the orthogonal direction to the encoder and reader planes (air gap separation), the requirements are more stringent. According to the reported data (Fig. 2.50a), during encoder motion, the guiding system must guarantee an air gap separation comprised between roughly 0.17 and 0.22 mm. Increasing the tolerance relative to the air gap separation is possible at the expense of larger encoder sizes. Nevertheless, the fabricated encoders have been correctly read with an in-house experimental setup. Consequently, the reported air gap tolerances seem to be reasonable for an eventual reader, based on a sufficiently robust mechanical guiding system, to be used for encoder reading in a real scenario.

References

1. Herrojo C, Paredes F, Mata-Contreras J, Ramon E, Núñez A, Martín F (2019) Time-domain signature barcodes: near-field chipless-RFID systems with high data capacity. IEEE Microwave Mag 20(12):87–101
2. Vena A, Perret E, Tedjini S (2011) Chipless RFID tag using hybrid coding technique. IEEE Trans Microw Theory Techn 59:3356–3364
3. Islam MA, Karmakar NC (2012) A novel compact printable dual-polarized chipless RFID system. IEEE Trans Microw Theory Techn 60:2142–2151
4. El-Awamry A, Khaliel M, Fawky A, El-Hadidy M, Kaiser T (2015) Novel notch modulation algorithm for enhancing the chipless RFID tags coding capacity. In: IEEE International Conference on RFID, San Diego, CA, USA, pp 25–31
5. https://www.gs1.org/standards/epc-rfid
6. Herrojo C, Mata-Contreras J, Paredes F, Martín F (2017) Near-field chipless RFID encoders with sequential bit reading and high data capacity. In: IEEE MTT-S International Microwave Symposium (IMS'17), Honolulu, Hawaii, June 2017
7. Herrojo C, Mata-Contreras J, Paredes F, Martín F (2017) Microwave encoders for chipless RFID and angular velocity sensors based on S-shaped split ring resonators (S-SRRs). IEEE Sensors J 17:4805–4813
8. Herrojo C, Mata-Contreras J, Paredes F, Núñez A, Ramón E, Martín F (2018) Near-field chipless-RFID tags with sequential bit reading implemented in plastic substrates. Int J Magnetism Magnetic Mat 459:322–327
9. Herrojo C, Mata-Contreras J, Paredes F, Martín F (2017) High data density and capacity in chipless radiofrequency identification (chipless-RFID) tags based on double-chains of S-shaped split ring resonators (S-SRRs). EPJ Appl Metamat 4(8):6 pages
10. Herrojo C, Mata-Contreras J, Paredes F, Martín F (2017) Near-field chipless RFID system with high data capacity for security and authentication applications. IEEE Trans Microw Theory Techn 65(12):5298–5308
11. Herrojo C, Mata-Contreras J, Paredes F, Núñez A, Ramon E, Martín F (2018) Near-field chipless-RFID system with erasable/programmable 40-bit tags inkjet printed on paper substrates. IEEE Microw Wireless Compon Lett 28(3):272–274
12. Herrojo C, Mata-Contreras J, Paredes F, Núñez A, Ramon E, Martín F (2018) Very low-cost 80-bit chipless-RFID tags inkjet printed on ordinary paper. Technologies 6(2):52
13. Herrojo C, Muela F, Mata-Contreras J, Paredes F, Martín F (2019) High-density microwave encoders for motion control and near-field chipless-RFID. IEEE Sensors J 19:3673–3682

14. Havlicek J, Herrojo C, Mata-Contreras J, Paredes F, Martín F (2018) Stub-loaded microstrip line loaded with half-wavelength resonators and application to near-field chipless-RFID. In: IEEE-MTT-S Latinoamerica Microwace Conf (LAMC'2018), Arequipa, Peru, Dec 2018
15. Havlicek J, Herrojo C, Paredes F, Martín F (2019) Enhancing the data density in near-field chipless-RFID systems with sequential bit reading. IEEE Antennas Wirel Propag Lett 18:89–92
16. Marques R, Medina F, Rafii-El-Idrissi R (2002) Role of bi-anisotropy in negative permeability and left handed metamaterials. Phys Rev B 65, paper 144441
17. Marqués R, Martín F, Sorolla M (2007) Metamaterials with negative parameters: theory design and microwave applications. John Wiley, Hoboken (NJ)
18. Martín F (2015) Artificial transmission lines for RF and microwave applications. John Wiley, Hoboken (NJ)
19. Schurig D, Mock JJ, Smith DR (2006) Electric-field-coupled resonators for negative permittivity metamaterials. Appl Phys Lett 88, paper 041109
20. Naqui J, Martín F (2013) Transmission lines loaded with bisymmetric resonators and their application to angular displacement and velocity sensors. IEEE Trans Microw Theory Tech 61(12):4700–4713
21. Chen H, Ran L, Huangfu J, Zhang X, Chen K, Grzegorczyk TM, Au Kong J (2004) Left-handed materials composed of only S-shaped resonators. Phys Rev E 70(5):057605
22. Chen H, Ran L, Huangfu J, Zhang X, Chen K, Grzegorczyk TM, Kong JA (2005) Negative refraction of a combined double S-shaped metamaterial. Appl Phys Lett 86(15):151909
23. Chen H, Ran L-X, Jiang Tao H-F, Zhang X-M, Cheng K-S, Grzegorczyk TM, Kong JA (2005) Magnetic properties of S-shaped split ring resonators. Prog Electromagn Res 51:231–247
24. Naqui J, Coromina J, Karami-Horestani A, Fumeaux C, Martín F (2015) Angular displacement and velocity sensors based on coplanar waveguides (CPWs) loaded with S-shaped split ring resonator (S-SRR). Sensors 15:9628–9650
25. Guo Y, Goussetis G, Feresidis AP, Vardaxoglou JC (2005) Efficient modeling of novel uniplanar left-handed metamaterials. IEEE Trans Microw Theory Tech 53:1462–1468
26. Pendry JB, Holden AJ, Robbins DJ, Stewart WJ (1999) Magnetism from conductors and enhanced nonlinear phenomena. IEEE Trans Microw Theory Tech 47:2075–2084
27. Martín F, Falcone F, Bonache J, Marqués R, Sorolla M (2003) Split ring resonator based left handed coplanar waveguide. Appl Phys Lett 83:4652–4654
28. Naqui J, Durán-Sindreu M, Martín F (2011) Novel sensors based on the symmetry properties of split ring resonators (SRRs). Sensors 11:7545–7553. https://doi.org/10.3390/s110807545
29. Naqui J, Martín F (2016) Application of broadside-coupled split ring resonator (BC-SRR) loaded transmission lines to the design of rotary encoders for space applications. In: IEEE MTT-S International Microwave Symposium (IMS'16), San Francisco, May 2016
30. Herrojo C, Paredes F, Mata-Contreras J, Zuffanelli S, Martín F (2017) Multistate multiresonator spectral signature barcodes implemented by means of S-shaped Split Ring Resonators (S-SRR). IEEE Trans Microw Theory Tech 65:2341–2352
31. Aznar F, Gil M, Bonache J, Baena JD, Jelinek L, Marqués R, Martín F (2008) Characterization of miniaturized metamaterial resonators coupled to planar transmission lines. J Appl Phys 104, paper 114501-1-8
32. Herrojo C, Paredes F, Martín F (2019) Double-stub loaded microstrip line reader for very high data density microwave encoders. IEEE Trans Microw Theory Techn 67(9):3527–3536

Chapter 3
System Requirements for Industrial Scenarios and Applications

In the previous chapter, the working principle and several examples of chipless-RFID systems based on time-domain signature barcodes were reported. It was shown that an unprecedented number of bits (only limited by tag size) is achievable with these systems, at the expense of tag reading by proximity (through near-field coupling with the reader) and proper alignment between the tag and the reader. It is remarkable the system based on the bandpass configuration of Sect. 2.3.2, exhibiting good tolerance against tag-reader distance and misalignments. Moreover, small-sized (60 mm $\times$ 34 mm and 60 mm $\times$ 6.4 mm) 100-bit tags based on linear (straight) metallic strips were reported in Sect. 2.3.3, although, in this case, tag reading requires a mechanical guiding system able to guarantee tag-reader positioning (separation and alignment) within more restrictive limits.

The time-domain signature barcodes presented in Chapter 2 were implemented in low-loss (but high-cost) microwave substrates and were fabricated by means of subtractive techniques (either milling or photo-etching) as a first step to validate system functionality. However, industrial needs force the use of low-cost tags implemented on flexible substrates (such as plastic substrates or paper) and manufactured by means of additive (printing) processes (e.g., rotogravure, flexography, screen-printing, inkjet, etc.). Moreover, for tag-cost reduction (a crucial aspect for the penetration of any chipless-RFID technology in the market), it is convenient to massively fabricate all-identical tags, by means of industrial manufacturing processes, and then personalize (program) such tags in a later stage. These aspects, intimately related to industrial and market requirements, are discussed in this chapter. Additionally, an improved version of the read-out electronics (as compared to the one used in Chapter 2), is also presented. The chapter also discusses an approach to read the tags synchronously, avoiding the need to use header bits to determine the tag velocity with regard to the reader, or even the need to implement the tags with equally spaced metallic elements.[1] The chapter ends with some prospective applications of

---

[1] Note that with synchronous tag reading, the relative velocity between the tag and the reader does not need to be constant.

© Springer Nature Switzerland AG 2020

F. Martín et al., *Time-Domain Signature Barcodes for Chipless-RFID and Sensing Applications*, Lecture Notes in Electrical Engineering 647,

https://doi.org/10.1007/978-3 030-39726-5_3

these time-domain and near-field chipless-RFID systems, including secure paper (the most canonical application), sensing, and motion control.

3.1 Tag Programming and Erasing

The idea of tag programming in chipless-RFID systems was pointed out in [1], where spectral signature barcodes (i.e., frequency-domain based tags) implemented with spiral resonators were considered. In such retransmission-based tags, where each spiral is tuned to a different frequency and provides one bit of information, the binary states ('0' or '1') are discerned by the absence or presence of a singularity (notch) in the frequency response of the tag. The absence of singularity (corresponding to the '0' state) can be generated, obviously, by the absence of the corresponding spiral resonator in the tag. Alternatively, as discussed in [1], resonator detuning, e.g., by short-circuiting the spirals, prevents from the appearance of the notch at the corresponding frequency, thereby achieving the same effect. This suggests that tags can be programmed after being fabricated by simply detuning certain resonant elements of the tag, making them inoperative. Resonator detuning can be achieved not only by short-circuiting it (an additive process), but also by cutting it (a subtractive process), providing a split. In the former case, inkjet printing may be used in order to add the necessary metallization (through conductive ink) to short-circuit the particle. In the second case, either milling or laser ablation are potential processes useful to detune the particle by fragmenting it. Figure 3.1 illustrates these programming processes and their effects on resonator detuning by considering an SRR coupled to a microstrip line.

For the chipless-RFID systems based on time-domain signature barcodes, where each resonant element provides a binary state (or bit), tags can be programmed following the procedures indicated in the previous paragraph. The main relevant advantage of tag programming is the fact that all-identical tags can be massively fabricated through industrial processes (hence reducing costs), and then programed

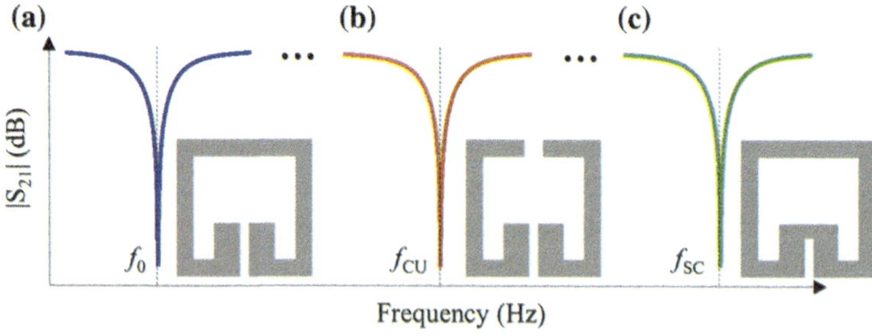

Fig. 3.1 Illustration showing tag programming. **a** Functional SRR; **b** detuned SRR achieved by cutting it; **c** and detuned SRR achieved by short-circuiting it. The fundamental resonance frequencies are indicated

in a later stage in order to provide the necessary ID code to each tag (or set of tags). By this means, a single mask (rather than as many masks as different ID codes) suffices for massively tag manufacturing, this being the key aspect to reduce tag production costs. According to the programming scheme indicated before, the mask pattern for tag production is the one corresponding to all-functional resonators. However, the mask providing all-inoperative (detuned) resonant elements can be also considered as an alternative. In this later case, all the resonators in the fabricated (all-identical) tags are detuned. Hence, tag programming requires either short-circuiting (if the fabricated detuned resonators are cut) or cutting (if the fabricated detuned resonators are short-circuited) those resonant elements that must provide the binary state '1' in the "written" tag. By this means, the functionality of such resonators is switched from detuned (as they are fabricated) to functional.

The tag programming scheme (short-circuiting or cutting the resonators) to be adopted in a real scenario should be determined by the cost of the considered process (inkjet printing, milling, laser ablation, etc.). For tag manufacturing with all-detuned resonators, the considered programming technique determines the layout of the resonant elements of the mask. To clarify this aspect, let us consider the resonant element of Fig. 3.1, the SRR, where resonator detuning is achieved with the indicated topologies (i.e., with a short-circuit or with a cut, with regard to the original functional resonator). If resonator cutting is the programming strategy, and fabricated tags with all-detuned resonators are desired, the SRRs of such massively fabricated tags (before programming) must exhibit a shorted-circuit, as shown in Fig. 3.1c. By contrast, the topology of Fig. 3.1b should be the considered one for the un-programmed tags, if tag programming by short-circuiting the resonant elements is the considered approach.

In this book, without loss of generality, tag programing is demonstrated by considering un-programed tags with all-functional elements (ID code '1111…') and by cutting the required resonant elements, in order to detune them and "write" the binary state '0' in such elements (equivalent to their absence). Indeed, the first programmed time-domain signature barcodes were those based on a double chain of S-SRRs [2], already reported in Sect. 2.3.1. It was demonstrated in that section that by providing a cut in the central position of the S-SRRs, these resonant elements are efficiently detuned (as revealed by the reported ID codes, where dips for such detuned S-SRRs are not present, see Fig. 2.19).

Tag programming was also demonstrated in [3] by cutting SRRs by the central position. The tags, identical to those reported in Sect. 2.3.2 and read by means of the SRR-loaded microstrip line also reported in that section, contain all the SRRs at their predefined positions, but some of them (the programmed ones) are detuned. Figure 3.2 shows the photograph of one of such detuned SRRs, in a 40-bit tag where the functional and detuned resonators alternate (ID code '101010…'). The measured normalized envelope function, also depicted in Fig. 3.2, confirms the validity of the programming approach, as revealed by the absence of dips associated to the inoperative SRRs (note that the tag response depicted in Fig. 3.2b is very similar to the one of the equivalent tag, namely, with identical ID code, depicted in Fig. 2.29).

The tags based on chains of linearly-shaped strips, described in Sect. 2.3.3, can be also programmed [4–6]. Indeed, the different reported ID codes were achieved

Fig. 3.2 Detail of a programmed 40-bit tag based on SRRs (**a**) and measured normalized envelope function (**b**). The cut SRRs alternate, providing the indicated ID code. SRR dimensions and tag substrate parameters are those reported in Sect. 2.3.2. Reprinted with permission from [3]; copyright 2017 IEEE

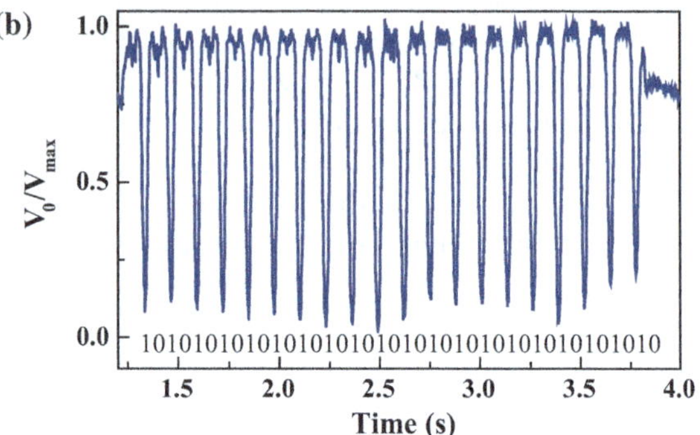

from a single fabricated tag, with all-functional strips, by slightly shortening the required strips (Sect. 2.3.3.1), or by cutting them by the central position (Sects. 2.3.3.2 and 2.3.3.3). Therefore, tag programing in these tags by a subtractive technique is effectively validated through the results presented in the previous chapter.

Tag erasing is the inverse process to tag programming (or writing), providing the original ID code (i.e., the one of the un-programmed tag, typically '1111…' or '0000…', as discussed before) to the erased tag. Obviously, tag erasing requires the "inverse" procedure to tag programming. Thus, if the tags are programmed by cutting certain resonators (or, more generally, by eliminating part of the metallization in the considered tag elements), it follows that it is necessary to metallize those previously subtracted regions in order to recover the original ID code (tag erasing). As long as an erased tag can be reprogrammed, it can be reused several times if needed.

The validation of these programmable/erasable time-domain signature chipless-RFID tags is demonstrated in [7, 8], where certain SRRs are first cut (tag programming), and then they are short-circuited by adding soldering tin [8], or conductive ink [7]. Figure 3.3 shows the photograph of a 14-bit (including four header bits) un-programmed tag. Two different ID codes have been generated by tag programming (resonator cutting) [8]. Such ID codes and the corresponding normalized envelope functions are also depicted in Fig. 3.3. Note that the header bits are used to adequately read the tags regardless of their relative orientation (face up or down) with regard to the reader (the header bits indicate if the bit sequence should be read from left to right or vice versa) [8]. Then, tag erasing has been carried out by short-circuiting the previously cut SRRs (see Fig. 3.4), and the resulting envelope function, also

Fig. 3.3 Photograph of the fabricated un-programmed 14-bit (including four header bits) tag with all code bits set to '1' (**a**) and measured normalized envelope function corresponding to the indicated programmed codes (**b**). The last code has been achieved after tag erasing, providing the code of the un-programmed tag. Reprinted with permission from [8]; copyright 2018 URSI

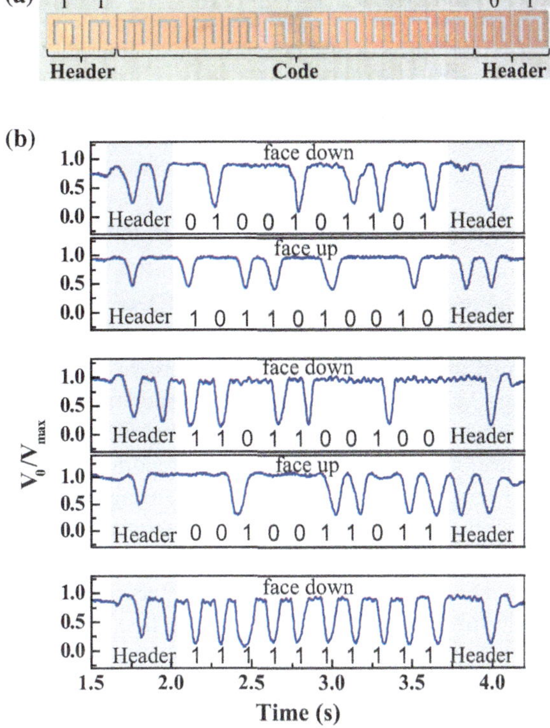

Fig. 3.4 Detail of a detuned (programmed) SRR (**a**) and SRR with soldering tin (erased) for functionalization (**b**)

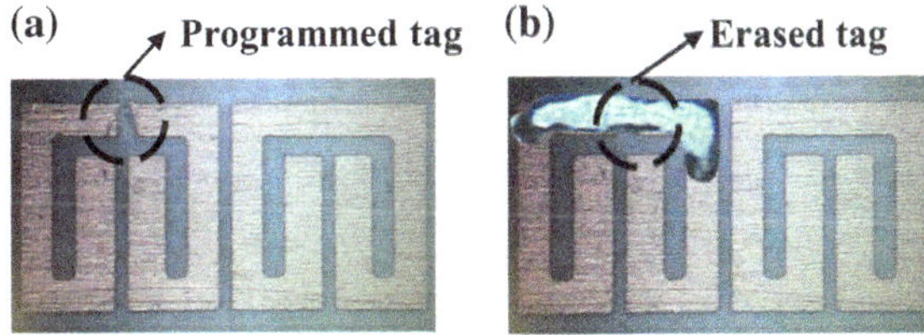

depicted in Fig. 3.3, reveals that the functionality of the resonant elements of the tag is recovered, hence validating tag erasing.

To end this section, it should be pointed out that the reported strategies for tag programming/erasing are consistent with tags where the different states associated to each resonant element are achieved by roughly preserving the topology of such resonators. This situation is typical in binary resonant elements, where the two logic states are differentiated by a subtle difference in their layouts (e.g., by a simple cut, as reported along this section). However, if the dimensions of the resonant elements of the tags are used to generate the different ID codes [9] or if the orientation of such elements is used for coding purposes [10] (typical encoding mechanisms in hybrid tags), then tag programming/erasing is not so simple, not to say impossible.

Therefore, as compared to hybrid tags (where a high data density can be achieved by virtue of the multiple bits associated to each resonant element), the time-domain signature chipless-RFID tags presented along this book can be easily programmed and erased. Additionally, such latter tags offer an unprecedented data capacity, only limited by tag size.

3.2   Implementation on Plastic and Paper Substrates

For the application of a chipless-RFID system in a real scenario, it is necessary to implement the tags on low-cost substrates, such as plastic or paper. This is an essential aspect in order to be competitive in terms of cost against chipped-RFID systems. Thus, in this section, the functionality of chipless-RFID systems based on time-domain signature barcodes implemented on plastic and paper substrates is demonstrated. The examples that are reported are based on tags where the ID codes have been generated by inkjet printing, using commercial conductive inks. Nevertheless, for massive tag fabrication in industrial processes, other printing techniques such as rotogravure, flexography, screen-printing, offset, etc., may result more efficient.

In [11, 12] time-domain signature barcodes based on S-SRRs and printed on a plastic substrate were reported for the first time. The reader is identical to the one reported in Sect. 2.3.1, based on an S-SRR-loaded CPW transmission line in stop-band configuration. The tags, with identical geometry to those of Sect. 2.3.1, contain 10 S-SRRs, hence providing 10 bits of information. The considered substrate for tag fabrication was the polyethylene naphtalate (PEN) film (*Dupont Teijin Q65FA*) with thickness $h = 125$ µm. The dielectric constant and loss tangent of this material was determined by means of the *Agilent 85072A* split cylinder resonator. The obtained values were found to be $\varepsilon_r = 3.36$ and $\tan\delta = 0.0042$. For tag printing, the *Ceradrop Ceraprinter X-Serie* inkjet printer, and two layers of *Dupont PE410* conductive ink (with conductivity 7.28×10^6 S/m) were used. The thickness of the printed layer was estimated to be 3.3–3.5 µm. Figure 3.5 depicts the photograph of a fabricated 10-bit tag, with all-functional S-SRRs, as well as the measured normalized envelope function of such tag, where 10 dips, associated to the resonant elements (all-functional), are visible. The frequency of the interrogation signal is identical to the one of Sect. 2.3.1, i.e., $f_c = 4$ GHz, since the PEN substrate is similar (in terms of thickness, dielectric constant, and loss tangent) to the low-loss microwave substrate used for tag fabrication in that section. Tag programming (discussed in Sect. 3.1) with tags printed on plastic substrates is also possible (e.g., by cutting certain S-SRRs by the central position). To validate the functionality of programmed tags, the ID code with '1's and '0's alternating was generated. Figure 3.6 shows a detail of a detuned resonator, as well as the measured normalized envelope function, where only 5 dips are visible. As it was discussed in Chapter 2, dip splitting is possible if the air gap and/or the frequency of the perfectly aligned S-SRRs (forming the BC-S-SRR particle) do vary during tag reading. Such (slight) splitting can be appreciated in the

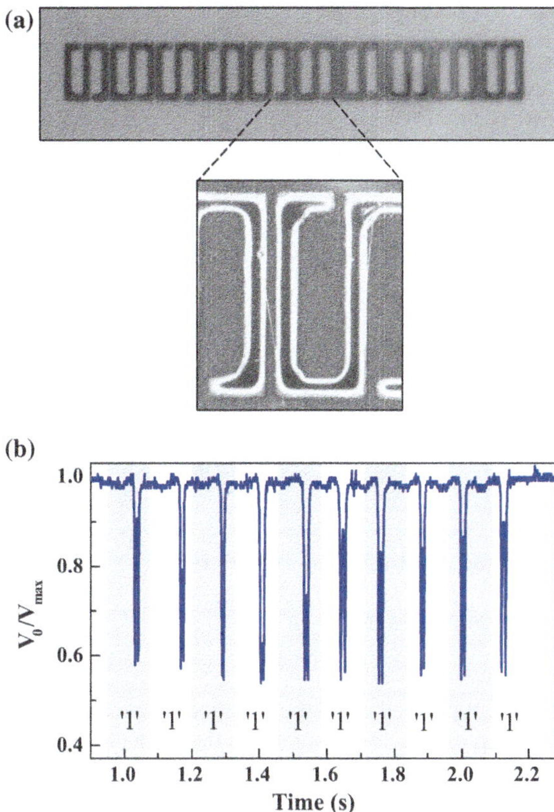

Fig. 3.5 Photograph of the fabricated 10-bit chipless-RFID tags, with code '1111111111' (**a**) and measured envelope function (**b**). Reprinted with permission from [12]; copyright 2018 Elsevier

figure. Nevertheless, this does not prevent from correct tag reading, provided such dip pairs (if present) are situated below a certain threshold value (e.g., 0.8).

The validity of PEN substrate for tag implementation by inkjet printing was also demonstrated by considering the system reported in Sect. 2.3.2, with a reader and tags based on SRRs [3]. As an illustrative result, Fig. 3.7 depicts the photograph of a fabricated 10-bit tag (with all-functional resonators) and the measured normalized envelope function (PEN substrate parameters, conductive ink and printing process are identical to those depicted in the preceding paragraph).

The use of paper substrate for the implementation of time-domain signature barcodes was first reported in [7], where the *PowerCoat HD ultra-smooth* paper, with thickness $h = 215$ μm, dielectric constant $\varepsilon_r = 3.11$ and loss tangent tanδ = 0.039, was used (the split cylinder resonator indicated above was used for the determination of ε_r and tanδ). In this case, a single layer of conductive ink (*Dupont PE410*) was printed on that substrate by means of the *Ceradrop Ceraprinter X-Serie* inkjet printer, providing an estimated measured thickness of 2.6 μm. A single 40-bit tag, based on SRRs, with two pairs of header bits was fabricated (Fig. 3.8). Then, such tag was programmed by cutting certain resonant elements (Fig. 3.9), providing the ID codes

Fig. 3.6 Photograph of the programmed 10-bit tag, with alternating detuned S-SRRs (**a**) and measured normalized envelope function (**b**). Reprinted with permission from [12]; copyright 2018 Elsevier

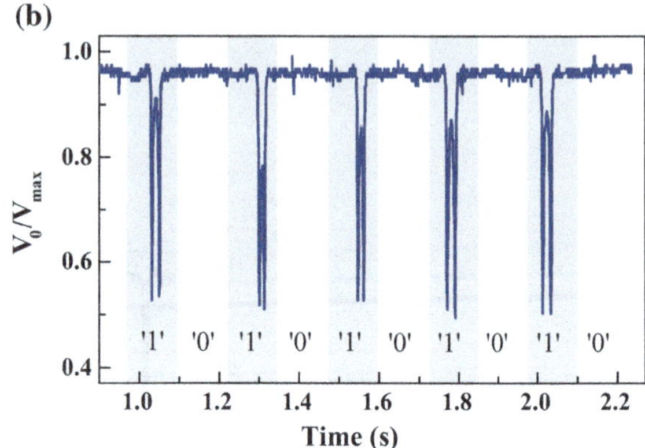

and the measured envelope functions depicted in Fig. 3.10 (obtained with the tags faced up or down with regard to the reader). Then, by adding conductive ink in the cut SRRs (Fig. 3.9), the tag was erased, reverting the functionality of the inoperative SRRs. The corresponding measured normalized envelopes, depicted in Fig. 3.11, reveal that the tag ID code is the canonical one, with all-functional resonators. Thus, these results point out that programmable/erasable tags printed on paper substrate can be correctly read, regardless of their orientation (faced up or down) with regard to the reader.

3.3 Improving the Read-Out Electronics

The chipless-RFID systems based on time-domain signature barcodes reported so far were validated by visualizing the ID codes in an oscilloscope and by generating the feeding (interrogation) signal by means of a commercial function generator. Moreover, the relative motion between the tag and the reader was achieved by means of a mechanical guiding system based on a step motor. In a real system, a tunable oscillator (e.g., based on a VCO) to generate the interrogation signal should be considered (however, this is the subject of many textbooks and is out of the scope of this book).

Concerning the read-out electronics, the use of the isolator (based on a circulator) to prevent mismatching reflections from the diode should be avoided. For that

Fig. 3.7 Photograph of the 10-bit printed chipless SRR based tag implemented on PEN substrate (**a**) and measured normalized envelope function (**b**). Reprinted with permission from [3]; copyright 2017 IEEE

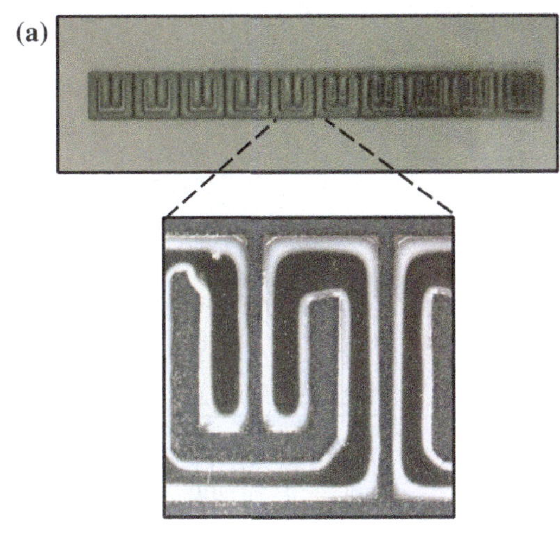

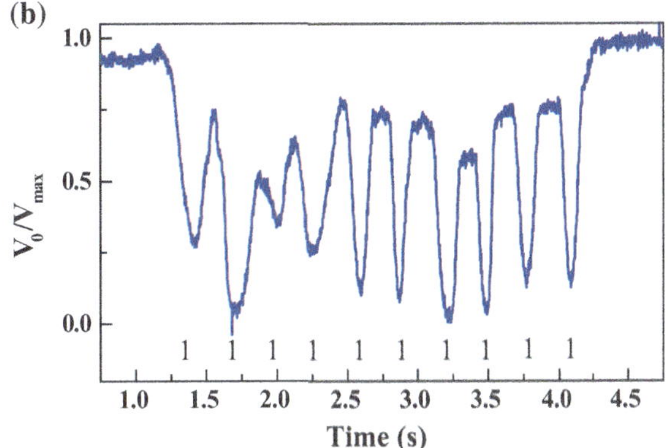

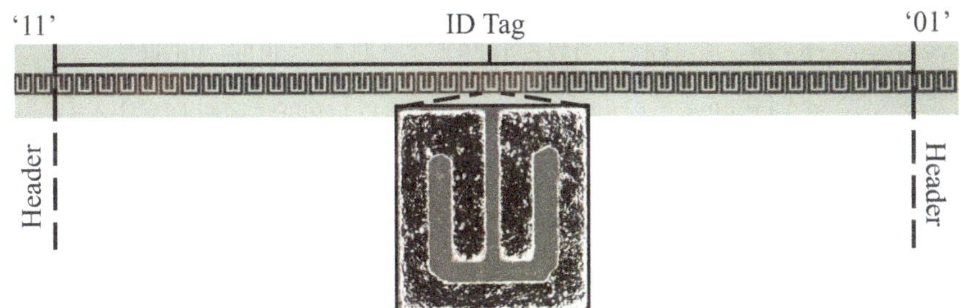

Fig. 3.8 Photograph of the fabricated 40-bit tag with header bits on paper substrate with all bits set to the logic state '1'

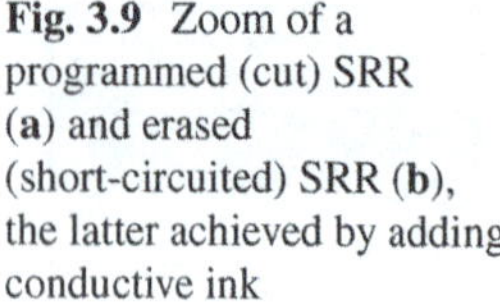
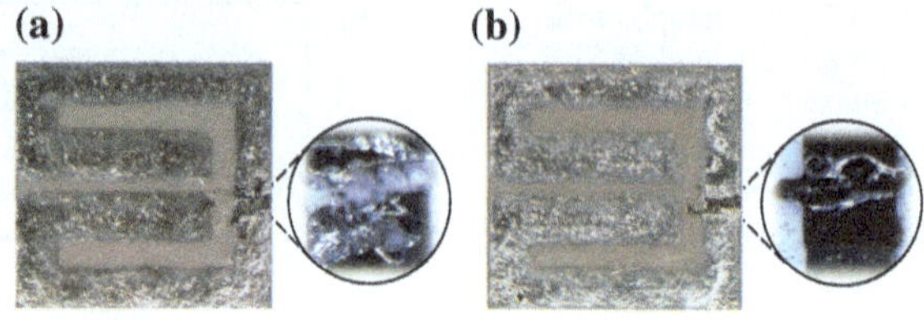

Fig. 3.9 Zoom of a programmed (cut) SRR (**a**) and erased (short-circuited) SRR (**b**), the latter achieved by adding conductive ink

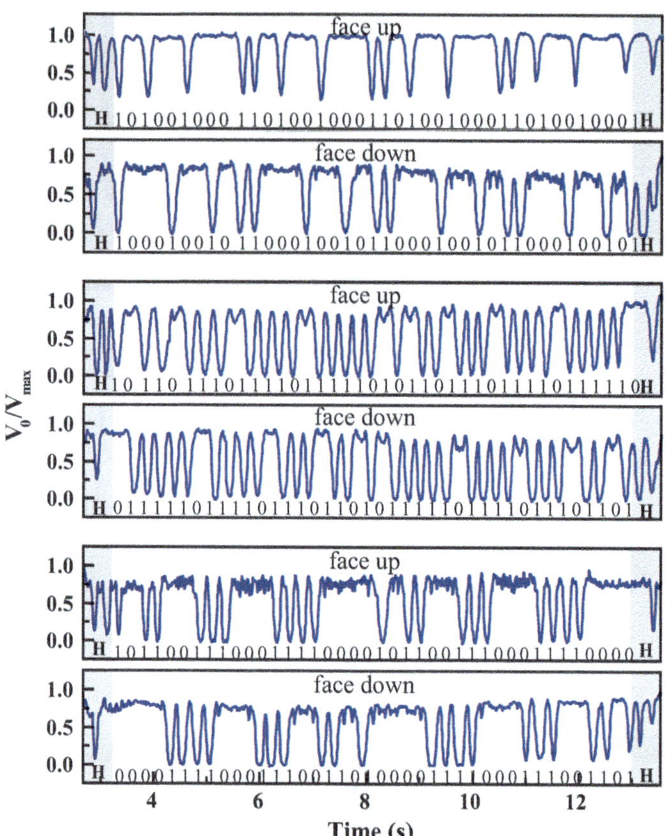

Fig. 3.10 Measured normalized envelope function for the 40-bit programmed tags with the indicated codes. Reprinted with permission from [7]; copyright 2018 IEEE

purpose, the envelope detector used in the systems reported so far, based on such isolator, plus a diode and an active probe, is replaced with an integrated circuit (*Analog Devices ADL5511*) able to provide the envelope function. This represents a cost reduction as long as the circulator, a high-cost device, is discarded. Moreover, rather than an oscilloscope, a real system demands for the use of a data acquisition card (e.g.,

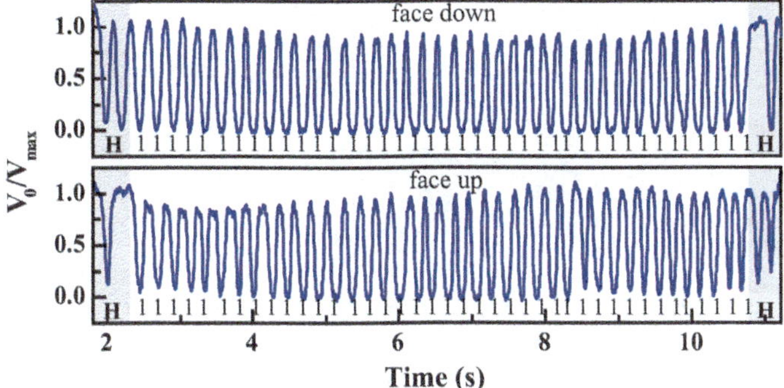

Fig. 3.11 Measured normalized envelope for the 40-bit erased tag. Reprinted with permission from [7]; copyright 2018 IEEE

the *National Instruments myRIO*), able to provide the sampled data to a computer. Then the ID codes can be automatically obtained by means of a post-processing unit.

Concerning the mechanical guiding system, a system similar to the one used to read credit cards in bank machines or bank notes (as can be encountered in many shops) may be envisaged. This mechanical part of the reader is out of the scope of this book. Nevertheless, a printer has been adapted and equipped with an *adhoc* guiding channel in order to absorb the tag. Such system guarantees good alignment and roughly constant air gap distance between the tag chain and the sensitive part of the reader, also conveniently placed and adapted to the modified printer. The system, including the adapted printer plus the improved readout electronics was validated by reading 80-bit tags (plus a pair of header bits) inkjet-printed on ordinary DIN A4 sized paper, to be discussed next.

3.4 System Validation

The validation of the chipless-RFID approach, including tag programming, tag functionality in ordinary paper, and tag reading through the scheme reported in the previous section is the subject of this section. Figure 3.12 depicts the photograph of the complete system. This setup was used in [13] to read 80-bit (plus the header bits) chipless-RFID tags inkjet-printed on ordinary (and low cost) DIN A4 paper, as mentioned before. The photograph of the printed tag, with all bits set to '1' is depicted in Fig. 3.13. Such tag was read by introducing it in the printer with the resonant elements of the tag faced up and down, producing the responses depicted in Fig. 3.14a. Then, tag programming was done by cutting (detuning) some resonant elements, writing the '0' logic state in the corresponding resonator. One of the resulting codes and its responses (faced up and down) are shown in Fig. 3.14b, where the functionality and validity of the system can be appreciated. These latter results, which represent a significant advance as compared to the first time-domain near-field chipless-RFID

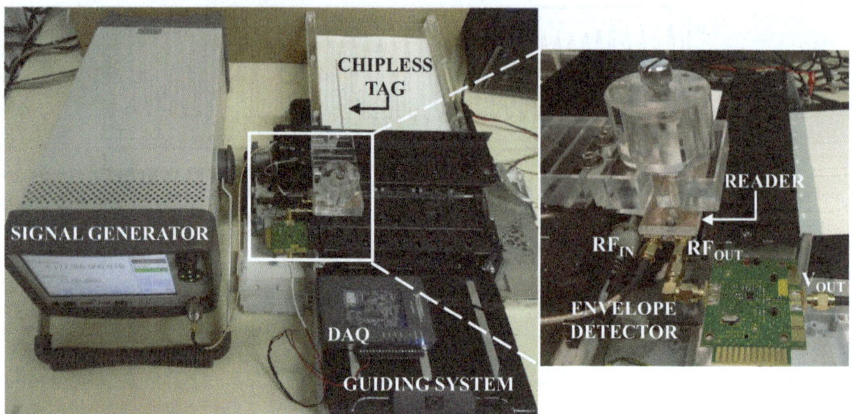

Fig. 3.12 Photograph of the set-up used for obtaining the tag responses (envelope functions) in those tags printed on DIN A4 paper

Fig. 3.13 Photograph of the 80-bit tag with header bits implemented on ordinary DIN A4 paper (with all bits set to the logic state '1'). The size of the tag is 3.35 mm × 281.9 mm

system prototype reported in Sect. 2.3.1 [14], are important since the functionality of the system is demonstrated by considering chipless-RFID tags printed on ordinary and low-cost paper.

3.5 Synchronous Tag Reading

In the chipless-RFID systems based on time-domain signature barcodes discussed so far it has been assumed that the distance between adjacent element positions (tuned or detuned) is uniform, and that the relative velocity between the tag and the reader is constant. Such velocity, if it is not known, can be obtained by adding additional bits (header bits) at the beginning of the tag chain. However, it cannot be always guaranteed that the relative displacement between the tag and the reader (necessary for tag reading) proceeds at uniform speed. If such velocity is not constant, bit reading cannot be carried out at periodic time intervals (by considering temporal windows), and synchronous reading is required. A simple scheme for synchronous reading consists of adding a chain of all-functional elements in the tag, with the elements

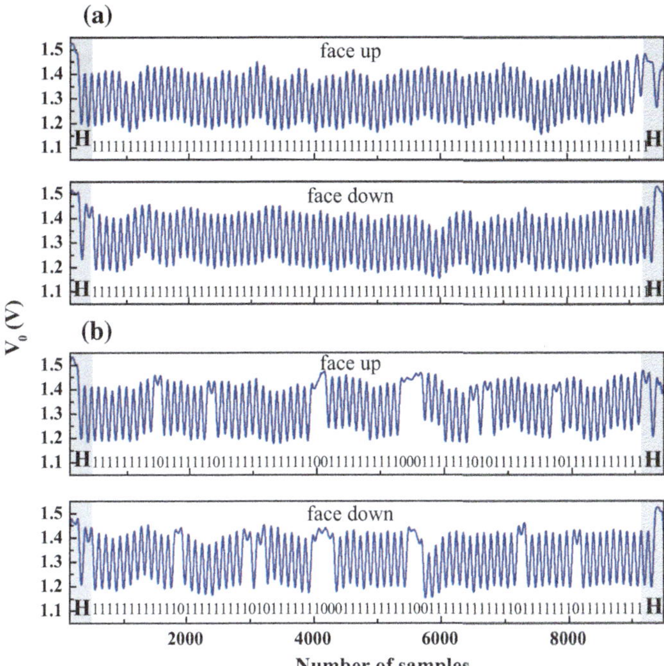

Fig. 3.14 Measured envelope of the inkjet-printed **a** 80-bit tag with all bits set to the logic state '1' and **b** 80-bit programmed tag with the indicated code and with header bits. Since the envelope function has been inferred from a data acquisition system in this validation experimental set-up, the abscise axis is the number of samples, rather than the time. Reprinted with permission from [13]; copyright 2018 MDPI

of this chain (clock chain) at the same positions of the elements (either tuned or detuned) of the other chain (ID chain). By using two independent readers (one for the clock chain and another one for the ID chain), the time intervals for reading the ID code are dictated by the information provided by the clock chain (revealed as peaks, or dips, in the corresponding envelope function). This synchronous reading scheme, illustrated in Fig. 3.15, is conceptually simple, but needs two independent readers and two element chains in the tag side.

An alternative approach for synchronous reading, based on a single tag chain and two sensing elements in the reader, is reported in this section [15]. One sensing element provides the ID code, whereas the other one is used for synchronization. Coding must be carried out in such a way that, for the ID sensing element, two states are distinguished, but not for the sensing element devoted to synchronization, that must detect the presence of tag elements regardless of their stored data ('0' or '1'). One potential (and preliminary) approach for achieving such functionality may be to implement the tags by means of linear chains of square-shaped or rectangular metallic patches, with tag encoding determined by patch dimensions. Namely, two different

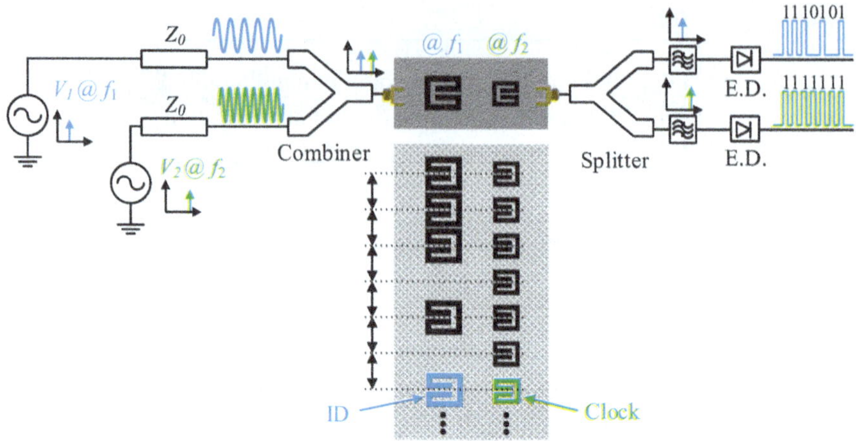

Fig. 3.15 Sketch of a synchronous reading system based on a pair of tag chains and two independent readers

patch dimensions must be considered, the larger patches providing one logic state and the smaller patches the other logic state. In the reader side, the information relative to the velocity (synchronization) and the information relative to the ID code can be independently inferred by considering, e.g., a microstrip transmission line loaded with two uncoupled and concentric complementary split ring resonators (CSRRs) tuned to different frequencies. The topology of the specific designed reader is depicted in Fig. 3.16a, whereas Fig. 3.16b shows the topology of a particular 5-bit tag.

In order to avoid inter-resonator coupling, both reader CSRRs (based on a single slot loop) have been designed with the slit located at the same angular position. The outer (third) slot ring has merely the function of tailoring the quality factor of the outer CSRR. For the smaller resonant element, the presence of a tag element on top of it modifies its resonance frequency regardless of its size. Therefore, this resonant element is the one used for synchronization purposes (clock). However, for the larger CSRR, its resonance frequency is only modified by the presence of the larger rectangular patches on top of it. Consequently, this CSRR is the one that determines the ID code. It should be emphasized that with this synchronous reading scheme, the distance between adjacent patches is not critical, and, indeed, tag periodicity is not actually required, contrary to the previous near-field chipless-RFID tags based on time-domain signature barcodes.

Figure 3.17 depicts the simulated frequency response of the bare reader, and the frequency responses of the reader loaded with a small and a large rectangular patch. Such simulations have been inferred by means of the *Keysight Momentum* commercial software. For the reader, the parameters of the *Rogers RO4003C* substrate with dielectric constant $\varepsilon_r = 3.38$, thickness $h = 1.52$ mm and loss factor $\tan\delta = 0.0022$, have been used; for the tag, the parameters of the *Rogers RO4003C* substrate with dielectric constant $\varepsilon_r = 3.38$, thickness $h = 0.2$ mm and loss factor

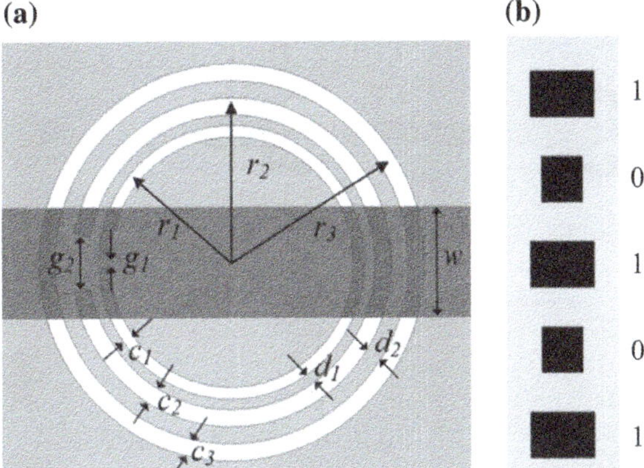

Fig. 3.16 Topology of the proposed reader (**a**) and 5-bit tag (**b**). Reader dimensions (in mm) are: $r_1 = 3.46$, $r_2 = 4.06$, $r_3 = 5$, $c_1 = 0.31$, $c_2 = 0.45$, $c_3 = 0.45$, $d_1 = 0.28$, $d_2 = 0.45$, $g_1 = 0.2$, $g_2 = 1.43$, $w = 2.9$ and the line length is 40 mm. For the tag, the width, as well as the separation between elements, is set to 8 mm, and the small and the large rectangular patch length are 7.2 mm and 11.5 mm, respectively. Note that the reader and the tag are not drawn to scale. The ID code of the tag is also indicated. Reprinted with permission from [15]; copyright 2019 IEEE

Fig. 3.17 Simulated frequency response (transmission coefficient) of the bare reader and reader loaded with small and large patch. Reprinted with permission from [15]; copyright 2019 IEEE

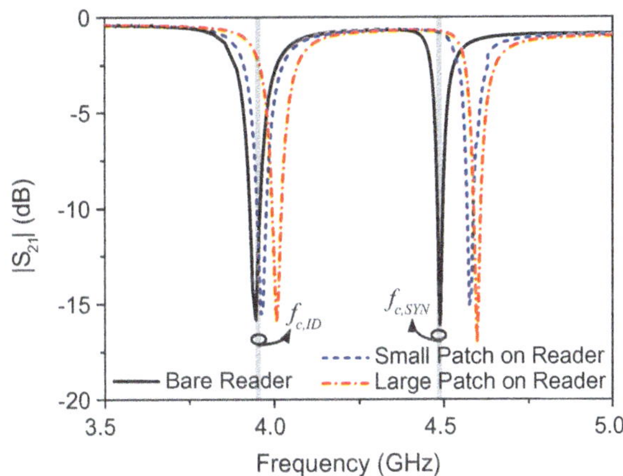

$\tan\delta = 0.0022$, have been considered. In the simulations, the vertical distance (air gap) between the tag and the reader has been considered to be 0.5 mm.

It can be seen from Fig. 3.17 that when both reader CSRRs are loaded with the small patch of the tag, only the larger resonance (the one of the smaller resonator) is appreciably shifted up. However, when the larger patch is on top of the sensing region of the reader, both resonances are notably shifted up. Thus, by tuning the harmonic interrogation signals to the frequencies indicated in Fig. 3.17, it is possible to infer the ID code (through the envelope function associated to the harmonic signal

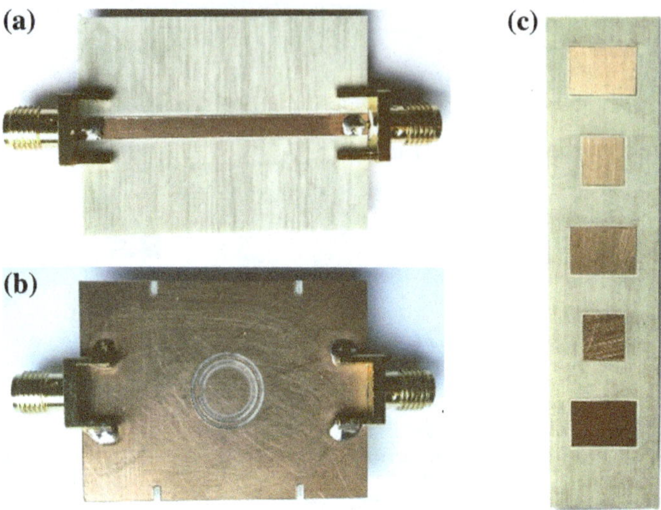

Fig. 3.18 Photograph of the reader: top (a); bottom view (b); and photograph of the 5-bit tag (c). Reprinted with permission from [15]; copyright 2019 IEEE

with smaller frequency) and the synchronization signal (given by the peaks in the envelope function associated to the harmonic signal with larger frequency).

The fabricated reader and the 5-bit tag are depicted in Fig. 3.18. For system validation, two harmonic signals are needed, as mentioned before, and such signals must be injected to the input port of the reader line (see Fig. 3.15). For that purpose a combiner may be used. Similarly, at the output port of the reader line, the information relative to the synchronization and the information relative to the ID code may be separately obtained by means of a diplexer (such diplexer would provide the corresponding envelope functions).[2] In this section, the main objective is to validate the proof-of-concept of the proposed synchronous scheme for tag reading. For this reason, rather than by using a combiner/diplexer scheme, validation is done by independently injecting the harmonic signals to the input port of the reader line (the corresponding envelope functions are recorded in each case).

The first harmonic signal is the one corresponding to the ID code (with $f_{c,ID}$ = 3.975 GHz). Before the measurement of the envelope function, the simulated transmission coefficient at $f_{c,ID}$ that results by displacing the tag over the reader, is obtained (see Fig. 3.19). The transmission coefficient is larger when the larger patches ('1' state) are on top of the resonant elements of the reader. The corresponding (measured) envelope function is depicted in Fig. 3.20, and it correlates with the transmission coefficient; namely, large and small peaks are obtained when the large and small patches, respectively, are on top of the sensitive part of the reader.

---

[2]Such combiner/diplexer scheme is also used in Chapter 4 to discern the velocity from the direction of motion in microwave rotary encoders. Interested readers are recommended to read Chapter 4.

Fig. 3.19 Simulated transmission coefficients at $f_{c,ID}$ and $f_{c,SYN}$ as the tag is displaced over the reader. Reprinted with permission from [15]; copyright 2019 IEEE

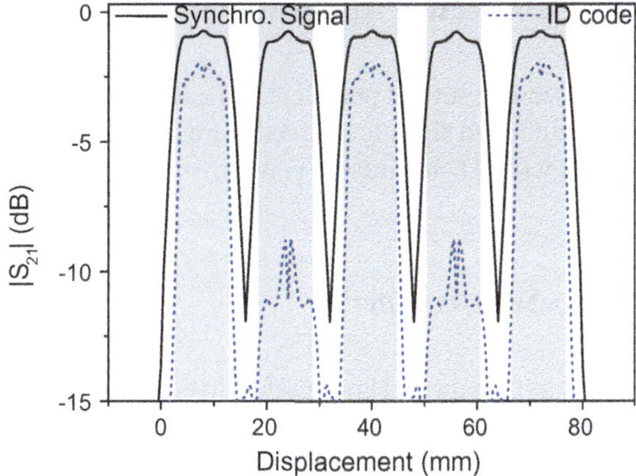

Fig. 3.20 Measured envelope functions corresponding to the reader synchronization signal and to the ID code. Reprinted with permission from [15]; copyright 2019 IEEE

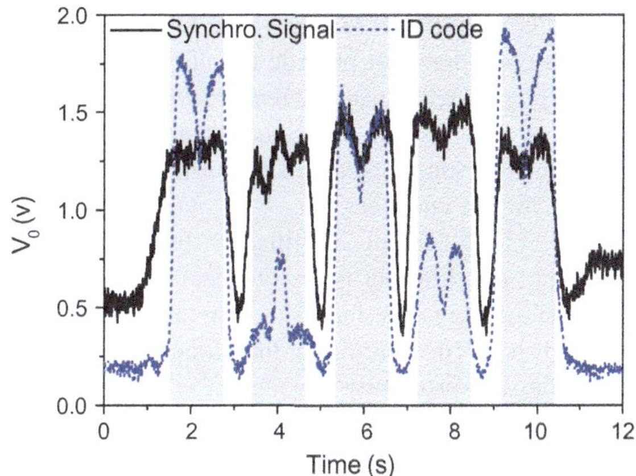

From this result, it is verified that tag reading correctly provides the ID code. The second harmonic signal, used to generate the synchronous signal, is tuned to $f_{c,SYN}$ = 4.520 GHz. The simulated transmission coefficient at $f_{c,SYN}$ is also depicted in Fig. 3.19. Note that identical peaks arise regardless of the size of the tag patches, indicating that a good synchronous signal is potentially achievable. To verify this, the measured envelope function is inferred (Fig. 3.20). It is corroborated from such function, that peaks of similar magnitude arise each time a patch (large or small) is present on top of the resonant elements of the reader. Thus, synchronous tag reading with this approach is possible. Although preliminary results have been reported, these results may constitute the seed for improved systems based on synchronous reading with a single tag chain.

3.6 Prospective Applications

In this section, several potential applications of the chipless-RFID technology based on time-domain signature barcodes are discussed, including secure paper, proximity sensors with ID, and motion control.

3.6.1 Secure Paper

As long as it has been demonstrated in the previous section that time-domain signature barcodes can be printed on ordinary paper and correctly read, it follows that the chipless-RFID system subject of this book can be of interest in applications devoted to secure paper (probably this is the most canonical application of these systems). By directly printing the ID codes on the document of interest, it is provided with a secure code, difficult to reproduce. Of course, the ID codes can be photocopied, but such copy does not provide the necessary characteristics of the original tag in order to be read with an adhoc reader system, similar to the one depicted in Fig. 3.12. Moreover, it is possible to fabricate the tags through lamination. In this latter case, the chain of resonant elements is buried, consequently increasing the difficulty in reproducing the encoders.

Corporate documents, certificates, banknotes, ballots, exams, medical prescriptions, etc., are different types of documents that can take benefit of the proposed time-domain signature barcodes, providing them further security against counterfeiting. It is worth mentioning that document identification and authentication (i.e., tag reading) by proximity (through near-field), as it is necessary in the system under study, may be beneficial in order to provide major levels of confidence against spying or eavesdropping, as compared to alternative systems based on far-field reading.

Some of the validation examples reported in the previous sections, especially those where paper substrate is used for tag printing, point out the potential of this technology in secure paper applications. Nevertheless, a further example is reported in this section, where a real application is envisaged: authentication of medical prescriptions. Such application may be of interest to fight against fraud in medicine dispensation, and one possible solution may be to print time-domain signature barcodes in the medical prescriptions, in order to identify the facultative physician authorized to prescribe the corresponding medicines. Figure 3.21 depicts the photograph of a Spanish medical prescription, where a 16-bit ID code has been printed through inkjet. Such code has then been correctly read through the system shown in Fig. 3.12 (the results are depicted in Fig. 3.22). It should be mentioned that the ID code has been generated by means of a single layer of conductive ink. Moreover, the quality of the paper used for medical prescription is very limited from an electromagnetic viewpoint. These facts strengthen the robustness of the proposed chipless-RFID system, where SRR-based time-domain signature barcodes for the proposed application (secure medical prescriptions) have been considered.

(a)

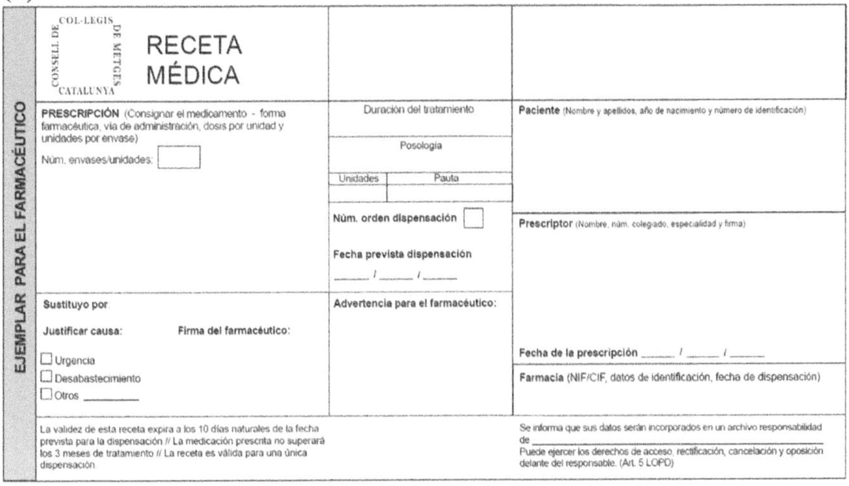

<table>
<tr><td rowspan="6" style="writing-mode:vertical-lr">EJEMPLAR PARA EL FARMACÉUTICO</td><td colspan="2">CONSELL DE COL·LEGIS DE METGES CATALUNYA
RECETA
MÉDICA</td><td></td><td></td></tr>
<tr><td>PRESCRIPCIÓN (Consignar el medicamento - forma farmacéutica, vía de administración, dosis por unidad y unidades por envase)

Núm. envases/unidades:</td><td>Duración del tratamiento

Posología

Unidades | Pauta</td><td>Paciente (Nombre y apellidos, año de nacimiento y número de identificación)</td></tr>
<tr><td></td><td>Núm. orden dispensación

Fecha prevista dispensación
_____ / _____ / _____</td><td>Prescriptor (Nombre, núm. colegiado, especialidad y firma)</td></tr>
<tr><td>Sustituyo por

Justificar causa: Firma del farmacéutico:

☐ Urgencia
☐ Desabastecimiento
☐ Otros _______</td><td>Advertencia para el farmacéutico:</td><td>Fecha de la prescripción _____ / _____ / _____

Farmacia (NIF/CIF, datos de identificación, fecha de dispensación)</td></tr>
<tr><td>La validez de esta receta expira a los 10 días naturales de la fecha prevista para la dispensación // La medicación prescrita no superará los 3 meses de tratamiento // La receta es válida para una única dispensación.</td><td></td><td>Se informa que sus datos serán incorporados en un archivo responsabilidad de _______________
Puede ejercer los derechos de acceso, rectificación, cancelación y oposición delante del responsable. (Art. 5 LOPD)</td></tr>
</table>

(b)

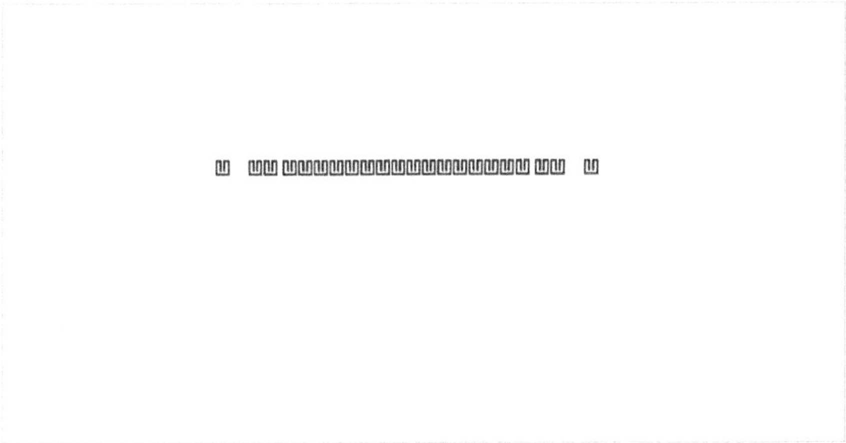

Fig. 3.21 Photograph of a Spanish medical prescription, **a** top view and **b** bottom view with a printed ID code

3.6.2 *Proximity Sensors with Identification*

In [16], a variant of the chipless-RFID system with near-field and sequential bit reading extensively reported and discussed in Chapter 2 was proposed. The difference concerns the reader and the reading mechanism, whereas the tags are identical to those described in Chapter 2 (i.e., chains of S-SRRs, SRRs, half-wavelength resonators, linear strips, etc.). In the systems reported in Chapter 2, tag reading was achieved by mechanically displacing the tags over the reader at short distance (hence sequentially determining the functionality of the tag resonators or inclusions). By contrast, in the

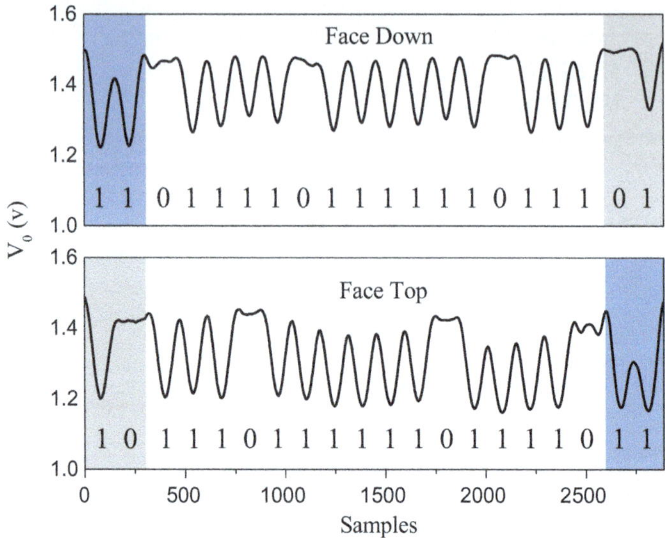

Fig. 3.22 Measured envelope function of the programmed inkjet-printed 16-bit tag with header bits. Since the envelope function has been inferred from a data acquisition system in this validation experimental set-up, the abscise axis is the number of samples, rather than the time

system described in [16], the tag (based on SRRs) merely needs to be positioned above the sensitive part of the reader (to be described), properly aligned with it. Moreover, the reader is based on an array of power splitters that feed as many lines (conveniently loaded) as resonant elements (or bits) of the tag, and tag reading proceeds sequentially by means of a switch.

Let us consider, without loss of generality, 4-bit tags based on SRRs, like those of Sect. 2.3.2. The sketch of the chipless-RFID system is depicted in Fig. 3.23. The interrogation signal, with frequency f_c, is injected into the input port of a microstrip line that divides the signal into as many channels as number of bits of the tag (four

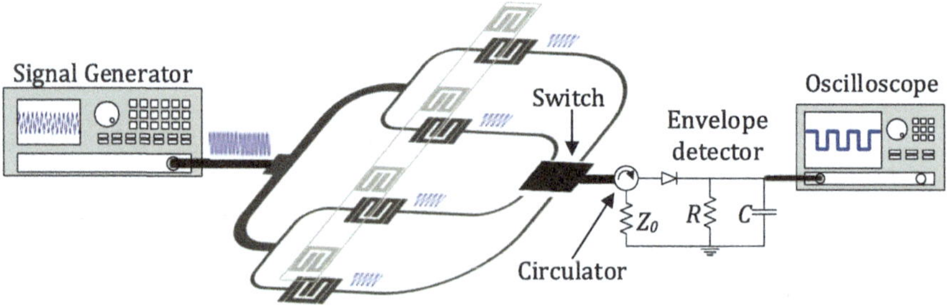

Fig. 3.23 Sketch of the proposed chipless-RFID sensing and identification system with switching reading. Reprinted with permission from [16]; copyright 2018 MDPI

in the case study). In each channel, a transmission line loaded with an SRR identical to those of the tag and oppositely oriented acts as sensitive part of the reader (the layout is the one depicted in Fig. 2.22a, except the dimensions). By positioning the tag on top of the reader (with the SRRs of the tag face-to-face to those of the reader, in close proximity), the electromagnetic coupling between the SRRs of the lines and the functional SRRs of the tag modifies the transmission coefficient of the corresponding lines. This effectively modifies (decreases) the amplitude of the feeding signal at the output port of such lines. Note that this modulation occurs only in those channel lines with a functional SRR on top of it. Therefore, the ID code is contained in the amplitudes of the output signals of the different channels, with high- and low-level amplitudes corresponding to the '0' and '1' logic states, respectively. As it was discussed in Sect. 2.3.2, the frequency of the feeding signal, f_c, must be selected in the bandpass of the frequency response of the unloaded channel lines (i.e., without SRRs on top of them). By this means, a significant decrease of the transmission coefficient at f_c when a functional SRR of the tag is placed on top of the SRR of the line is obtained, hence providing a high amplitude contrast between the '0' and '1' logic states. This aspect is important in order to increase the vertical detection distance (air gap).

As shown in Fig. 3.23, tag reading proceeds sequentially by means of a switch, with its output port connected to an envelope detector which provides the amplitude of the output port of each channel line, and hence the ID code of the tag. Such codes can only be detected (read) if the tag is within the influence of the electromagnetic fields generated by the channel line (a distance of the order of 1 mm). Therefore, the proposed system can be used as proximity detector with identification functionality. In other words, it is possible to identify the presence of a target in a certain position (very close to the reader or detector) and identify it.

The layout of the channel lines and SRR (tag) dimensions were optimized in [16] for system operation at $f_c = 2.25$ GHz. The active part of the reader was implemented on the *Rogers RO3010* substrate with thickness $h = 0.635$ mm and dielectric constant $\varepsilon_r = 10.2$. The first 4-bit tags were implemented on the same substrate. Figure 3.24a shows the photograph of the active part of the reader with a tag with all-functional resonators on top of it. The simulated and measured frequency responses between port 1 and 2 for the unloaded reader and for the reader loaded with the tag, considering different air gaps, are shown in Fig. 3.24b, c, respectively. It can be appreciated that by tuning the feeding signal in the vicinity of 2.25 GHz, the excursion experienced by the transmission coefficient is significant up to air gaps close to 1 mm. In terms of dynamic range, the optimum situation corresponds to an air gap of 0.5 mm since for this case the transmission zero is roughly situated at 2.25 GHz, and the transmission coefficient experiences a variation of the order of 40 dB. It should be mentioned that the responses of Fig. 3.24 correspond to the transmission coefficient between the input port and the output of port 2. Nevertheless, similar responses are obtained for the transmission coefficient of the other output ports, as demonstrated in [16].

As mentioned before, tag reading is carried out by means of a switching scheme, where each reader channel is sequentially selected. The corresponding signal is driven to the output port of the switch, connected to the input port of the envelope detector.

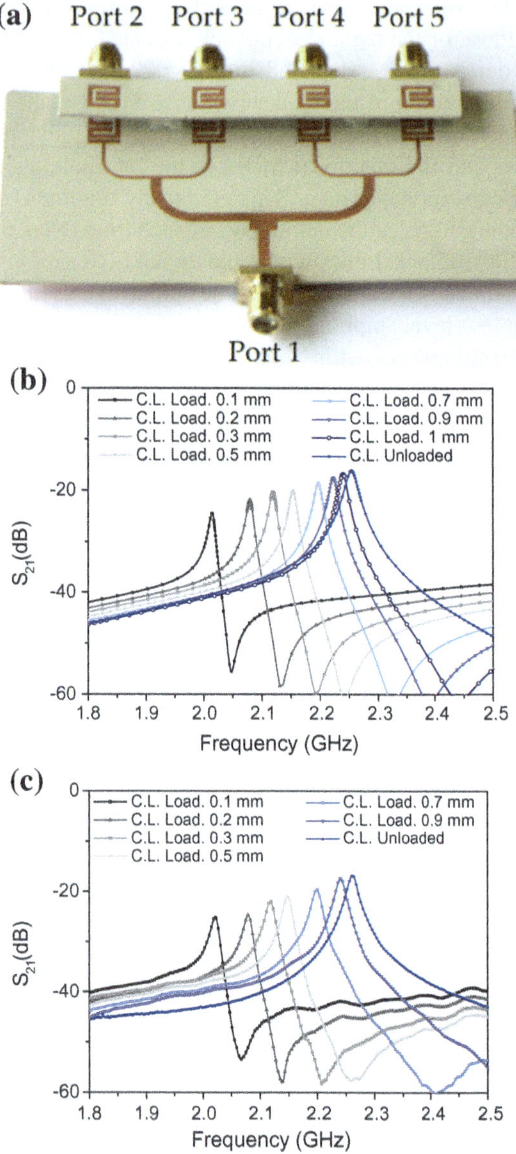

Fig. 3.24 Frequency responses of the channel line without tag on top of it and with tag SRR (perfectly aligned) at different distances. **a** Photograph of the prototype with tag on top and faced up; **b** electromagnetic simulation; and **c** measurement. Dimensions (in reference to Fig. 2.22a) are $W_1 = 0.58$ mm, $W_2 = 0.87$ mm, $l_1 = 5.53$ mm, $l_2 = 5.86$ mm, $s_1 = 0.35$ mm, $s_2 = 0.2$ mm. Reprinted with permission from [16]; copyright 2018 MDPI

The switch is based on the *Analog Devices HMC241AQS16E* integrated circuit. The photograph of the switch circuit is depicted in Fig. 3.25a. The switch requires two supply pins (VCC and GND), as well as pins A and B to select the switch input channels. In order to select the switch output ports automatically, and to manage the switching times, an *Atmel ATmega328P* microcontroller and the necessary electronic components, disposed in a printed circuit board (PCB), were use (Fig. 3.25b).

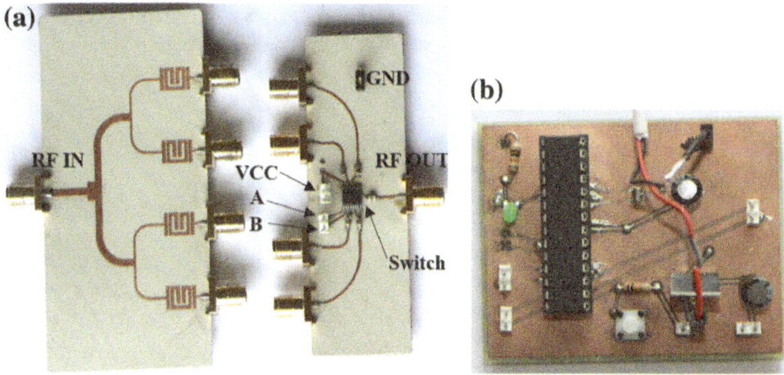

Fig. 3.25 **a** Photograph of the power divider and switch and **b** Photograph of the control circuit. The switch is managed by means of the microcontroller, which is responsible for carrying out a sequential scan of the channel lines as well as powering the switch. Reprinted with permission from [16]; copyright 2018 MDPI

The experimental setup used for system validation is shown in Fig. 3.26. The *Agilent E44338C* signal generator was used to feed the power divider with a harmonic signal (whose amplitude is reduced 6 dB at each channel line of the power divider). The switch, managed by the microcontroller and powered by an universal serial bus (USB) power tank, sequentially redirects the RF input channel lines to an RF output. To be able to discern the channel line position, the microcontroller was programmed to wait 0.1 s between channels sweep, and the switch time was configured to take 2 s at the first channel line and 1 s for the rest of the lines (see Fig. 3.27, where the gray

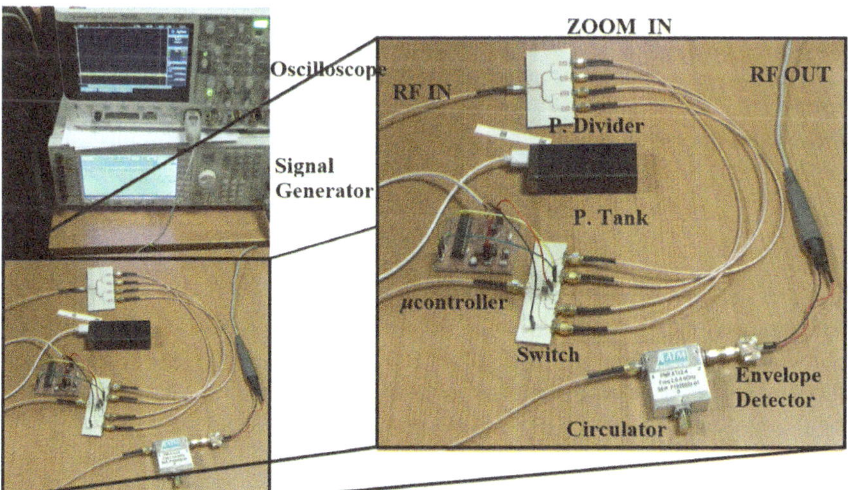

Fig. 3.26 Photograph of the experimental setup, where all components are shown. Reprinted with permission from [16]; copyright 2018 MDPI

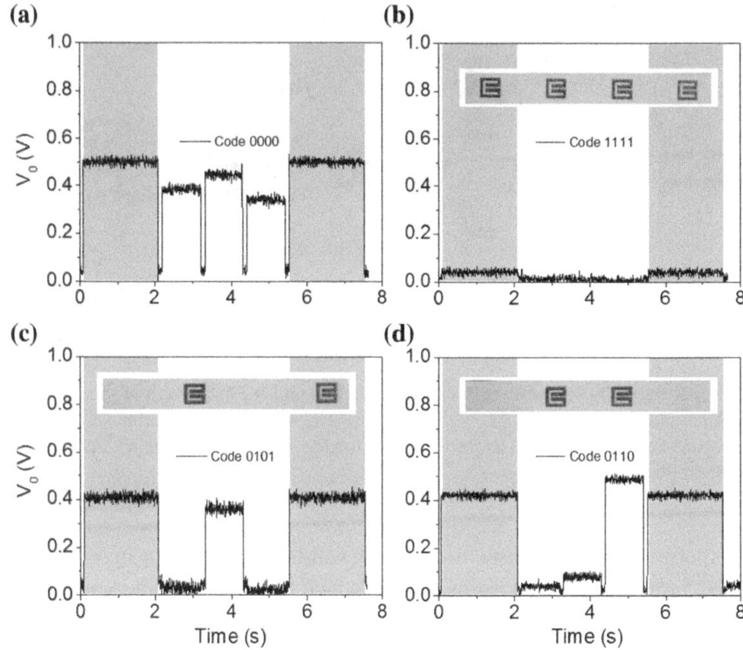

Fig. 3.27 Envelope functions of the tags with the indicated codes. The considered codes are: (a) '0000'; (b) '1111'; (c) '0101'; (d) '0110'. These responses were obtained by leaving the tag to rest on top of the reader lines. The gray regions show the first channel line, where the switch time was configured to 2 s. Reprinted with permission from [16]; copyright 2018 MDPI

regions indicate the 2 s time of the first channel). By this means, the initiation of the bit sequence is clearly identified. The switch RF output is connected to the envelope detector, which is indeed preceded by an isolator (implemented by means of the *L3 Narda-ATM ATc4-8* circulator), in order to prevent from unwanted reflections from the diode, as discussed in Chapter 2. The diode used to obtain the envelope function was the *Avago HSMS-2860*, whereas the necessary low-pass filter was implemented by means of the *Agilent N2795A* active probe (with resistance and capacitance $R = 1$ MΩ and $C = 1$ pF, respectively), connected to an oscilloscope (model *Agilent MSO-X-3104A*).

In order to validate the system by using low-cost tags, a 4-bit chipless-RFID tag was fabricated by using one layer of *Dupont^{TM} PE410* Ag conductive ink (with thickness 2.6 μm and conductivity 7.28×10^6 S/m) inkjet printed on *PowerCoat^{TM} HD ultra-smooth* paper with dielectric constant $\varepsilon_r = 3.1$ and thickness $h = 230$ μm (the *Ceradrop Ceraprinter X-Serie* inkjet printer was used). Switching reading was first applied in absence of tag, equivalent to a tag without resonators (with code '0000'). From this response (see Fig. 3.27a), it is reasonable to set the threshold level to differentiate the two logic states to 0.2 V. Further tag readings, corresponding to

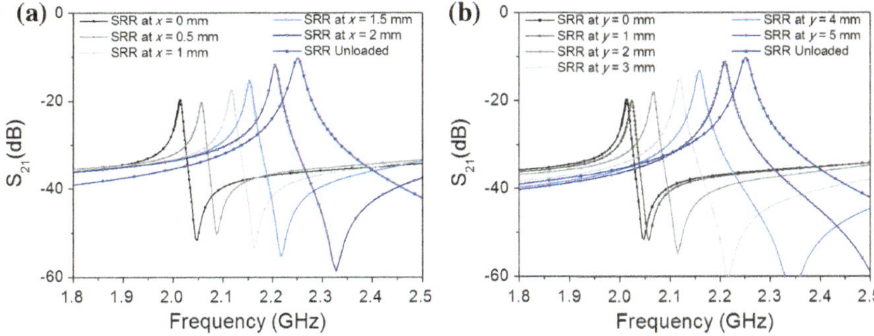

Fig. 3.28 Electromagnetic simulation of the transmission coefficient for the channel line with SRR on top of it and the indicated longitudinal (**a**) and lateral (**b**) misalignments. Reprinted with permission from [16]; copyright 2018 MDPI

different codes with SRRs faced up, are included in Fig. 3.27. The obtained responses validate the proposed approach.

One aspect that may limit the correct tag reading is the longitudinal and lateral misalignment between the SRRs of the tag and those of the reader. The effects of such misalignments were studied by electromagnetic simulation in [16]. Thus, Fig. 3.28 depicts the simulated responses of one SRR over one channel line for different longitudinal and lateral displacements (with the air gap set to 0.1 mm). As it can be seen, a significant excursion of the transmission coefficient is maintained up to 2 mm misalignment in the x-direction (longitudinal), and up to 5 mm misalignment in the y-direction (lateral). Therefore, these results point out that the proposed system is quite robust against lateral and longitudinal misalignments between the tag and the reader.

To end this subsection, the authors would like to mention that, as compared to the near-field chipless-RFID systems reported in Chapter 2, the main advantage of the near-field chipless-RFID system based on switching reading is the fact that a mechanical displacement system for the tag is not needed. Nevertheless, scaling up this system to many bits is difficult, and therefore the system should be focused on applications requiring a limited number of bits.

3.6.3 Motion Control

The chipless-RFID systems studied and discussed in Chapter 2 can be easily adapted to the implementation of displacement and velocity sensors. Since tag reading is carried out by a relative displacement between the tag and the reader, the measurement of such displacement and velocity is simple. For that purpose, the tag should be fabricated with all-functional resonators (corresponding to all bits set to '1'). Thus, if the distance between resonant elements is known, the cumulative number of peaks

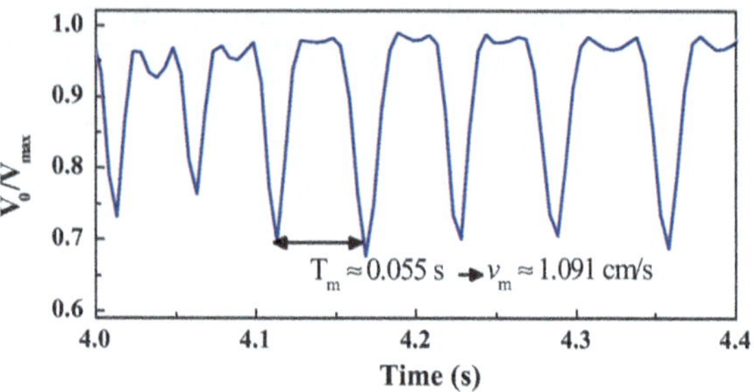

Fig. 3.29 Measured normalized envelope function obtained by displacing the strip chain of Fig. 2.35, with all strips functional, over the reader (only a portion of the curve is shown for a better view). Reprinted with permission from [4]; copyright 2019 IEEE

(or dips) in the envelope function determines the relative displacement (from a given reference position) between the tag and the reader. Moreover, from the time distance between adjacent peaks, or dips, the instantaneous relative velocity between the tag and the reader can be also inferred.

The dynamic range for the measurement of linear displacements is given by the length of the tag, whereas the spatial resolution is determined by the chain period. One advantage of this system, as compared to other linear displacement sensors based on resonator loaded lines [17–23], is the achievable combination of dynamic range and spatial resolution. By using the tags reported in Sects. 2.3.3.1, 2.3.3.2 or 2.3.3.3 [4–6], the spatial resolution is as good as 0.6 mm (the chain period). The dynamic range is 6 cm, since 100 linear strips were used in those tags, but this value can be enhanced by simply extending the length of the tag (hence accommodating more linear strips).

Figure 3.29 shows the envelope function of the tag depicted in Fig. 2.35a, and read by means of the structure depicted in Fig. 2.35b. All the strips in that tag are functional, and therefore the instantaneous linear velocity is simply given by the distance between adjacent dips (appearing when the intermediate position between two strips is perfectly aligned with the line axis of the reader). Such dip times can be easily recorded by means of a post-processing unit. From them, the linear velocity and the relative displacement between the reader and the tag chain from the reference position can be determined. According to Fig. 3.29, the time interval between adjacent dips is 55 ms, corresponding to a displacement velocity of 1.09 cm/s.

References

1. Preradovic S, Balbin I, Karmakar NC, Swiegers GF (2009) Multiresonator-based chipless RFID system for low-cost item tracking. IEEE Trans Microw Theory Techn 57:1411–1419
2. Herrojo C, Mata-Contreras J, Paredes F, Martín F (2017) High data density and capacity in chipless radiofrequency identification (chipless-RFID) tags based on double-chains of s-shaped split ring resonators (S-SRRs). *EPJ Appl Metamat* 4: 8–6
3. Herrojo C, Mata-Contreras J, Paredes F, Martín F (2017) Near-field chipless RFID system with high data capacity for security and authentication applications. IEEE Trans Microw Theory Techn 65(12):5298–5308

4. Herrojo C, Muela F, Mata-Contreras J, Paredes F, Martín F (2019) High-density microwave encoders for motion control and near-field chipless-RFID. IEEE Sens J 19:3673–3682
5. Havlicek J, Herrojo C, Mata-Contreras J, Paredes F, Martín F (2018) Stub-loaded microstrip line loaded with half-wavelength resonators and application to near-field chipless-RFID. In: IEEE-MTT-S LatinoAmerica microwace conference (LAMC'2018). Arequipa, Peru, Dec. 2018
6. Havlicek J, Herrojo C, Paredes F, Martín F (2019) Enhancing the data density in near-field chipless-RFID systems with sequential bit reading. *IEEE Antennas Wirel Propag Lett* 18: 89–92
7. Herrojo C, Mata-Contreras J, Paredes F, Núñez A, Ramon E, Martín F (2018) Near-field chipless-RFID system with erasable/programmable 40-bit tags inkjet printed on paper substrates. IEEE Microw Wireless Compon Lett 28:272–274
8. Herrojo C, Moras M, Paredes F, Mata-Contreras J, Núñez A, Ramon E, Martín F (2018) Erasable/programmable chipless-RFID tags with orientation-independent sequential bit reading. In: 2018 Spanish URSI conference, Granada, Spain, Sept 2018
9. Vena A, Perret E, Tedjini S (2011) Chipless RFID tag using hybrid coding technique. IEEE Trans Microw Theory Techn 59:3356–3364
10. Herrojo C, Paredes F, Mata-Contreras J, Zuffanelli S, Martín F (2017) Multi-state multi-resonator spectral signature barcodes implemented by means of s-shaped split ring resonators (S-SRR). IEEE Trans Microw Theory Techn 65(7):2341–2352
11. Herrojo C, Mata-Contreras J, Paredes F, Núñez A, Ramón E, Martín F (2017) Near-field chipless-RFID tags with sequential bit reading implemented in plastic substrates. In: Moscow international symposium on magnetism (MISM'17), Moscow, Russia, 1–5 July 2017
12. Herrojo C, Mata-Contreras J, Paredes F, Núñez A, Ramón E, Martín F (2018) Near-field chipless-RFID tags with sequential bit reading implemented in plastic substrates. J Magn Magn Mater 459:322–327
13. Herrojo C, Mata-Contreras J, Paredes F, Núñez A, Ramon E, Martín F (2018) Very low-cost 80-bit chipless-RFID tags inkjet printed on ordinary paper. Technol 6(2):52
14. Herrojo C, Mata-Contreras J, Paredes F, Martín F (2017) Near-field chipless RFID encoders with sequential bit reading and high data capacity. In: IEEE MTT-S international microwave symposium (IMS'17), Honolulu, Hawaii, June 2017
15. Paredes F, Herrojo C, Martin F (2019) An approach for synchronous reading of near-field chipless-RFID tags. In: 10th IEEE international conference on RFID technology and applications (IEEE RFID-TA 2019), Pisa, Italy, Sep 2019, pp 25–27
16. Paredes F, Herrojo C, Mata-Contreras J, Moras M, Núñez A, Ramon E, Martín F (2018) Near-field chipless-RFID sensing and identification system with switching reading. Sens 18:1148
17. Mandel C, Kubina B, Schüßler M, Jakoby R (2011) Passive chipless wireless sensor for two-dimensional displacement measurement. In: Proceedings in 41st European microwave conference, Manchester, UK, 2011, pp. 79–82
18. Naqui J, Durán-Sindreu M, Martín F (2011) Novel sensors based on the symmetry properties of split ring resonators (SRRs). Sens 11:7545–7553
19. Naqui J, Durán-Sindreu M, Martín F (2012) Alignment and position sensors based on split ring resonators. Sens 12(9):11790–11797
20. Karami-Horestani A, Fumeaux C, Al-Sarawi SF, Abbott D (2013) Displacement sensor based on diamond-shaped tapered split ring resonator. IEEE Sens J 13(4):1153–1160
21. Horestani A, Abbott D, Fumeaux C (2013) Rotation sensor based on horn-shaped split ring resonator. IEEE Sens J 13(8):3014–3015
22. Horestani AK, Naqui J, Abbott D, Fumeaux C, Martín F (2014) Two-dimensional displacement and alignment sensor based on reflection coefficients of open microstrip lines loaded with split ring resonators. Electron Lett 50(8):620–622
23. Ebrahimi A, Withayachumnankul W, Al-Sarawi S, Abbott D (2014) Metamaterial-inspired rotation sensor with wide dynamic range. IEEE Sens J 14(8):2609–2614

Chapter 4
Microwave Rotary Encoders

It was shown in Chap. 3 that the relative position and instantaneous velocity between the tag and the reader in a chipless-RFID system based on time-domain signature barcodes can be easily inferred by measuring the time distance between dips, or pulses, in the envelope function. If the system is intended to be applied to motion control (i.e., as linear displacement and velocity sensor), the ID code of the tag must be well known, so that from the space between functional resonant elements or strips in the tag, the tag position and velocity can be determined. As it was discussed in Sect. 3.6.3, it is convenient to implement the tag with all-functional inclusions (code '111...'), since the spatial resolution (intimately related to the number of pulses in the envelope function) is optimized by considering such ID code. In the present chapter, it is shown that by conveniently arranging the resonators of the tag forming a circular chain, the system can be used as a microwave rotary encoder.[1] Thus, the angular displacement and velocity of the circular resonator chain (printed or etched in the rotor), relative to the sensitive part of the reader (which acts as stator), can be recorded. These microwave rotary encoders may represent a good alternative to optical rotary encoders, especially in applications where operation in extreme environments (e.g., with high levels of pollution, dirtiness or radiation) is required. Moreover, microwave rotary encoders are implemented by means of low-cost elements. Therefore, in terms of cost, such encoders are competitive in front of optical rotary encoders or other encoders based on magnetic elements (e.g., Hall-effect encoders).

4.1 Working Principle

The working principle of the microwave rotary encoders is indeed the same as the one governing chipless-RFID systems based on time-domain signature barcodes. The main difference concerns the fact that the resonant elements of the tag are necessarily

---

[1] In the proposed rotary encoders of this chapter, the tag inclusions are resonant elements.

© Springer Nature Switzerland AG 2020

F. Martín et al., *Time-Domain Signature Barcodes for Chipless-RFID and Sensing Applications*, Lecture Notes in Electrical Engineering 647,
https://doi.org/10.1007/978-3-030-39726-5_4

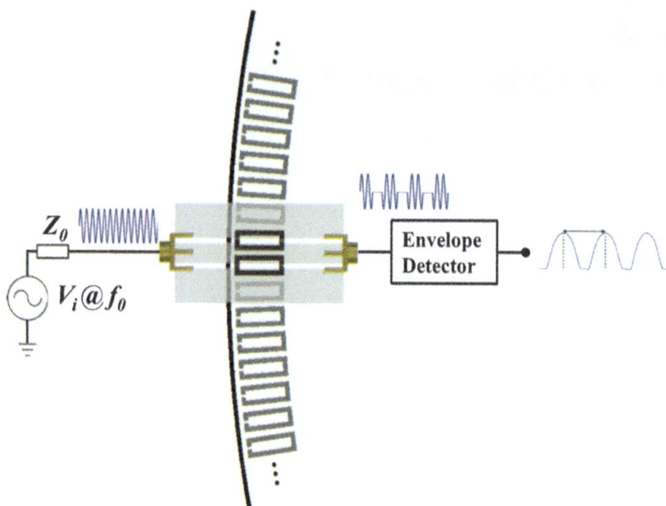

Fig. 4.1 Sketch of the microwave rotary encoder. Reprinted with permission from [2]; copyright 2017 IEEE

disposed forming a circular chain, printed or etched in the external perimeter of the rotor [1]. Within the context of microwave rotary encoders, rather than tag and reader, the more appropriate nomenclature to designate the key elements is *rotor* and *stator*. The rotor refers to the rotating element, which contains the chain of resonant elements; the stator is the fixed element, which generates the necessary signal (a harmonic signal) to determine the angular displacement and velocity of the rotor with regard to the stator. Figure 4.1 depicts a sketch of the microwave rotary encoder, showing the operating principle [2]. By rotor motion, the electromagnetic coupling between the transmission line of the stator (a CPW transmission line in Fig. 4.1) and the resonant elements of the rotor (SRRs in Fig. 4.1) is modulated, and the output signal is amplitude modulated, provided it is conveniently tuned, as discussed in Sect. 2.1. From the distance between adjacent maxima or minima in the envelope function, the angular velocity is inferred, since the angular step corresponding to adjacent resonators is known. The cumulative number of dips, or pulses, provides the angular displacement.

4.2 Comparative Analysis with Other Angular Velocity Sensors

A figure of merit in rotary encoders is the number of pulses per revolution, *PPR*, with direct impact on angle resolution and on the capability to detect rotation speed variations in short time intervals. Optical encoders with high values of *PPR*, reaching thousands of pulses in some cases, can be designed. Such encoders typically use a

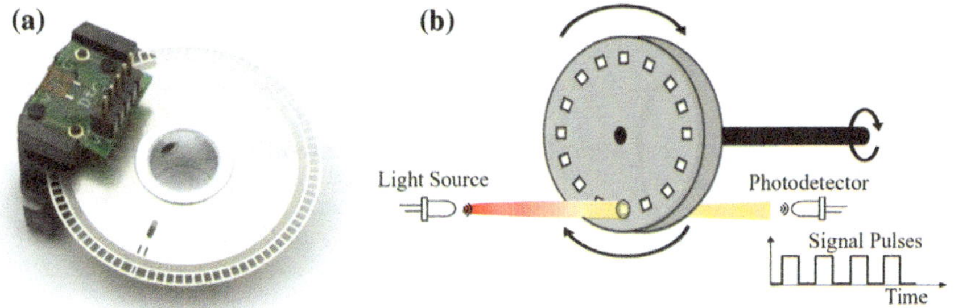

Fig. 4.2 Photograph (**a**) of an optical encoders and sketch (**b**) showing the working principle of such encoders

light source impinging a photo detector through slits in a metal or glass disc (Fig. 4.2). As compared to optical rotary encoders, microwave encoders are not competitive in terms of sensor performance (e.g., angular resolution) due to the lower achievable *PPR* values. However, as anticipated before, microwave encoders are cheaper and can be used in scenarios with harsh environmental conditions, such as space, as an alternative to optical encoders (which may suffer from radiation effects) [2]. Also, microwave encoders are more robust (as compared to optical encoders) against contaminants such as dust, dirt, liquids, and grease. Note, however, that the working principle of optical and microwave encoders is very similar. The main difference concerns the excitation signal, optical or electromagnetic, used to detect the presence of discontinuities in the rotor (i.e., slits in optical systems and printed resonators in microwave systems).

As compared to other angular velocity sensors based on microwave technology, the sensors analyzed in this chapter are very competitive in terms of *PPR* since, as will be shown, up to 1.200 pulses per revolution have been demonstrated [2]. For instance, in [3–5], angular velocity sensors based on a principle similar to the one reported in the previous section were reported. However, in those sensors, a single resonant element (an ELC resonator [6]), in axial configuration, was used (Fig. 4.3). The circular ELC resonator rotates axial to the rotor on top of a CPW transmission line acting as stator. It can be noted that, given a certain relative angular displacement between the rotor and the stator, the coupling between the ELC resonator and the CPW is identical to the one corresponding to the same orientation rotated 180°. Consequently, only two pulses per revolution (i.e., $PPR = 2$) are obtained with this sensing approach. Such sensors cannot be used to measure instantaneous angular velocities, but just uniform velocities or average values. As angular displacement sensors, microwave encoders offer competitive angle resolution (i.e., $\Delta\varphi = 360°/PPR$), and are robust to external electromagnetic interference (EMI) or noise as compared to the sensors proposed in [3–5], or to the sensors proposed in [7], based on the variations in the notch depth magnitude. That is, in the sensors reported in [3–5, 7], the angle is given by the variation of the notch depth experienced when the resonator element rotates. The sensor in [7] offers good sensitivity (variation of the notch depth with the angle), but the input dynamic range is very limited. By contrast, in the sensors reported in [3–5],

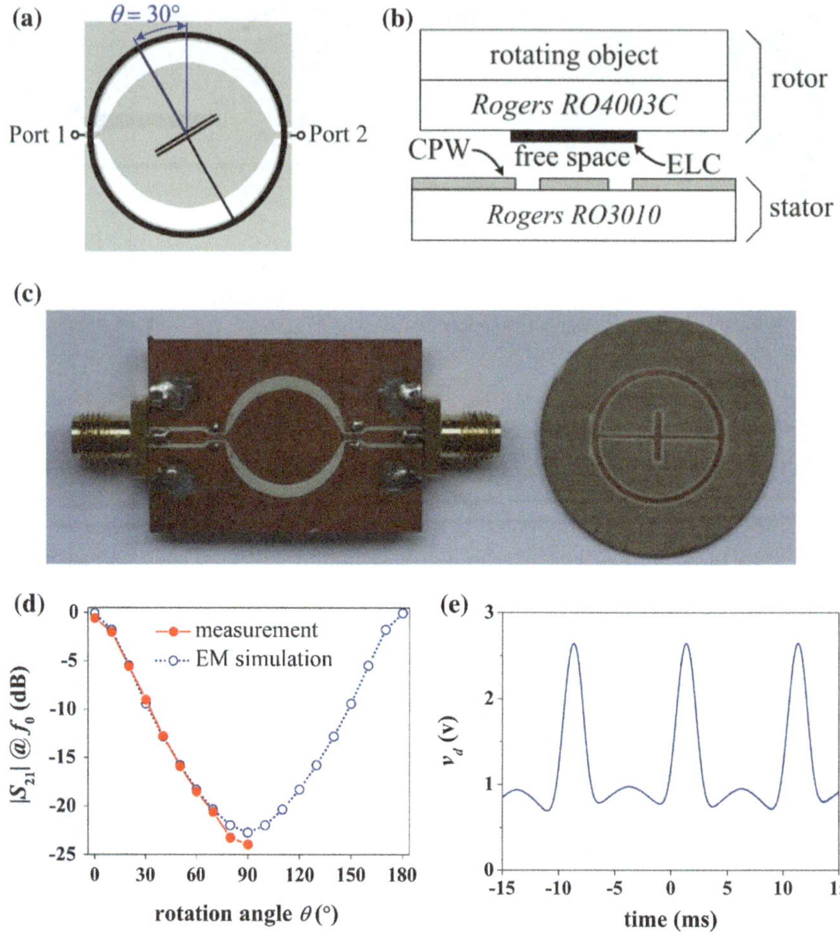

Fig. 4.3 Angular displacement and velocity sensor based on a ELC resonator in axial configuration. **a** Topology of the stator and rotor; **b** cross sectional view; **c** photograph of the stator (circularly shaped CPW transmission line) and rotor (circular ELC resonator); **d** variation of the transmission coefficient at the resonance frequency of the ELC resonator, used to determine angular displacement; **e** envelope function, exhibiting two pulses per revolution, used to determine the average angular velocity (the nominal value being 50 cycles per second). Reprinted with permission from [3]; copyright 2013 IEEE

the input dynamic range is 90°. In the sensors subject of this chapter, the number of pulses directly gives the angle so that calibration or linearization is not necessary and the dynamic range is unlimited.

In the sensors reported in [8], the angular displacement and velocity measurements are based on a wireless link between two circularly polarized antennas, and the relative angular displacement between both antennas is determined from the phase inferred from the frequency modulated signal generated in the receiver antenna. This system offers good angle resolution ($< 1.5°$), but the velocity is indirectly measured from the angle. Moreover the complete system may be complex in certain applications due to the orientation requirements of the antennas. In the microwave rotary encoders presented in this chapter, the angle resolution has been demonstrated to be $0.3°$ (as will be shown), angular velocity is directly given by the time distance between pulses, and the system is simple since antennas are not required and it is based on an AM modulation scheme.

4.3 Rotor and Stator Design

In this section, two different systems of angular displacement and velocity sensors, with rotors based on circular chains of SRRs etched along the edge, are considered. In one configuration, the stator is a SRR-loaded CPW transmission line; in the other sensing system, the stator consists of a SRR-loaded microstrip line. Such configurations are discussed next, and then they are compared.

4.3.1 Systems Based on Stator Implemented in CPW Technology

The first microwave rotary encoders based on chains of resonant elements located along the edge of the rotor (edge configuration) were reported in [1], and then such encoders were exhaustively studied in [2].

4.3.1.1 Rotor with a Single Chain of SRRs

In the first implementation, a single circular chain of SRRs was etched in the rotor, whereas the stator consisted of a CPW transmission line loaded with a pair of SRRs, etched in the backside of the substrate. The dimensions and separation between the SRRs of the stator were considered to be identical to those of the rotor. The cross sectional view of the sensor layers and the topologies of the stator and rotor (zoom view) are depicted in Fig. 4.4. It can be appreciated that the SRRs of the stator and those of the rotor are oppositely oriented for the same reason explained in Sect. 2.2 in regard to the oppositely oriented S-SRRs of tag and reader in the chipless-RFID systems discussed in that section. By these means, by adequately tuning the feeding signal of the stator, inter-resonator coupling in the rotor and multiple couplings between the stator and the rotor are circumvented.

When a pair of SRRs in the rotor is perfectly aligned (face to face) with the SRR pair of the stator, the resulting vertical pairs of SRRs are coupled to the line, but such

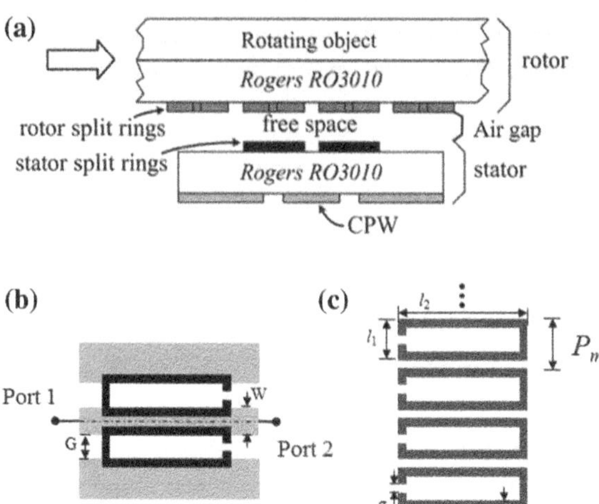

Fig. 4.4 Sensor cross-section (**a**), and layout of the (**b**) stator and **c** rotor. Dimensions are (in mm): $W = 1.3$, $G = 0.9$, $P_m = 2$, $l_1 = 1.6$, $l_2 = 6.2$, $c = 0.4$, and $g = 0.2$. The substrates have relative permittivity of 11.2, thickness of 0.635 mm (stator) and 1.27 mm (rotor), and loss tangent of 0.0023. Note that although the SRRs of the rotor are disposed in a circular chain, their geometry is rectangular. Since the radius of the rotor is much larger than SRR dimensions, it can be considered that the space between SRRs is uniform and that they can form perfectly broadside geometries with the SRRs of the stator

coupling occurs at the resonance frequency of the vertical (coupled) pairs, substantially different from the resonance frequency of a single split ring. Such vertically coupled SRRs, with rings rotated 180°, constitute the broadside coupled split ring resonator (BC-SRR) [9, 10] (see Sect. 2.3.2.1 for further details). As mentioned in that section, the key advantage of this configuration is the fact that since the coupling between the line and the BC-SRRs occurs at a frequency substantially lower than the frequency of the single SRRs, the above cited extra couplings are avoided by tuning the harmonic feeding signal at the resonance frequency of the BC-SRRs (this strategy was first introduced in [1]). The insertion loss of the CPW with a pair of SRRs of the rotor perfectly aligned with those of the stator (forming a perfectly aligned BC-SRR) is depicted in Fig. 4.5. Besides the response for this relative position between the rotor and the stator (called reference, REF, position), with perfectly aligned SRRs, the figure includes the response for other relative incremental/decremental displacements (corresponding to misalignments), expressed in terms of the period P_m of the rotor.

According to Fig. 4.5, the attenuation at f_0, the frequency of the perfectly aligned BC-SRRs, strongly depends on the relative displacement between the rotor and the stator. However, there is not a firm reason, a priori, to tune the feeding signal at such frequency. Indeed, if we chose a frequency different from f_0, two attenuation peaks, rather than one, are expected as the rotor moves one period (between $-P_m/2$ and $+ P_m/2$) above the stator (thereby doubling the *PPR*). As closer the selected frequency

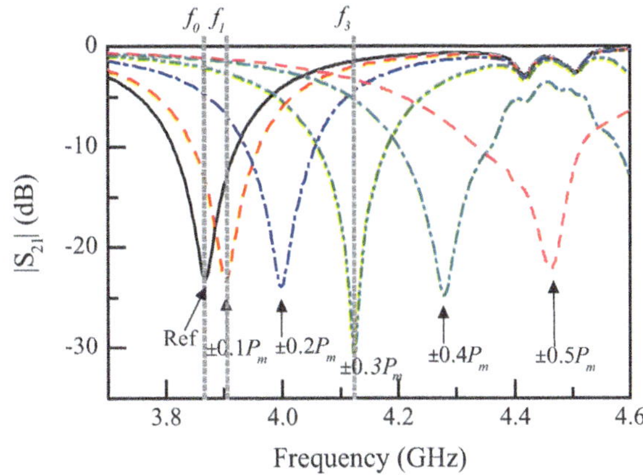

Fig. 4.5 Electromagnetic simulation, using *Keysight Momentum*, of the transmission coefficient for different relative displacements between the stator and the rotor (air gap of 0.5 mm). Reprinted with permission from [2]; copyright 2017 IEEE

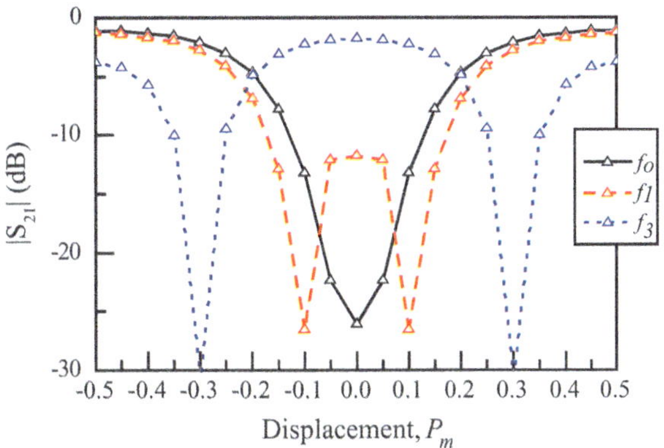

Fig. 4.6 Attenuation as a function of the rotor displacement at the indicated frequencies. Reprinted with permission from [2]; copyright 2017 IEEE

to f_0 is, the smaller the separation between attenuation peaks becomes. This effect is illustrated in Fig. 4.6, where the insertion loss for different frequencies, as the rotor is displaced above the stator, is depicted. Note also that as the two attenuation peaks approximate (by driving the frequency towards f_0), the transmission level between such peaks decreases, and the peaks merge when the considered frequency is f_0. The main conclusion of this study is that in order to increase the number of pulses per cycle, it is convenient to feed the CPW transmission line with a carrier frequency f_c satisfying $f_c \neq f_0$. The specific frequency value should be chosen in order to obtain similar transmission between attenuation peaks (e.g., $f_c \approx f_3$ is a good choice, according to Fig. 4.6). By injecting a harmonic signal tuned to that frequency, the rotor motion will modulate the amplitude at the output port of the CPW transmission line. From the envelope function, variations in the instantaneous rotation speed can be inferred with as much precision as higher the number of pulses per cycle is. With the proposed strategy, the number of pulses per cycle is twice the number of pulses achieved in the first implementation (reported in [1]). It is important to mention that if the harmonic feeding signal is tuned close to the resonance frequency corresponding

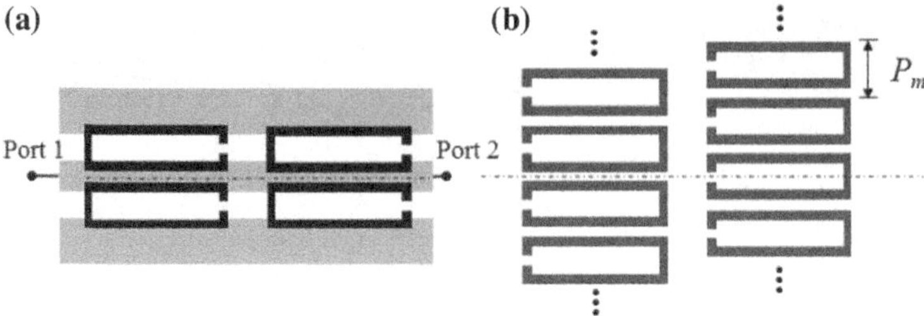

Fig. 4.7 Stator (**a**) and zoom view of the rotor (**b**) for the case of rotor with two SRR chains

to a relative displacement of a semi-period ($\pm\, 0.5P_m$ misalignment), the modulation index will be degraded. This is clear on account of Fig. 4.5, where it can be seen that the notch depth corresponding to the above-mentioned misalignment between the rotor and the stator is somehow smaller. For the robustness of the sensor, the highest possible notch depth is desired, since unwanted (but unavoidable) effects such as air gap variations, vibrations, etc., tend to reduce the modulation index of the AM signal at the output port of the stator.[2]

4.3.1.2 Rotor with Two Chains of SRRs for Resolution Improvement

To increase the number of pulses per revolution, an alternative approach is to consider two circular SRR chains in the rotor and two pairs of SRRs in the stator, as depicted in Fig. 4.7. The centers of the split rings of one chain are located in the azimuthal positions corresponding to the intermediate separation points of the split rings of the other chain.

Figure 4.8 depicts the response (insertion loss) of the CPW transmission line (stator) for different relative displacements between stator and rotor. The behavior is similar to the one of Fig. 4.5, in reference to the case with a single SRR chain. However, a rich phenomenology, to be discussed next, arises by the presence of two SRR chains. First of all, if the frequency of the feeding signal is the resonance frequency of the perfectly aligned BC-SRR, f_0, two attenuation peaks, rather than one, appear when the rotor experiences a relative motion between $-P_m/2$ and $+P_m/2$ (see Fig. 4.9). The reason is that there are two positions within this range where either the inner or the outer SRR chain forms a perfectly aligned BC-SRR with one of the SRRs of the stator. For frequencies different from f_0, four attenuation peaks, rather than two, are expected as the rotor moves within one SRR chain period, as can be seen in Fig. 4.9. Indeed, each SRR chain acts independently, and for that reason the number of attenuation peaks is twice the one corresponding to a single chain of SRRs. There is, however, a special case. Namely, if the frequency of the feeding signal is chosen as the one corresponding to the resonance for a relative

---

[2] To automatically determine the pulses, or dips, in the envelope function, it is convenient to maximize the variation experienced by such function.

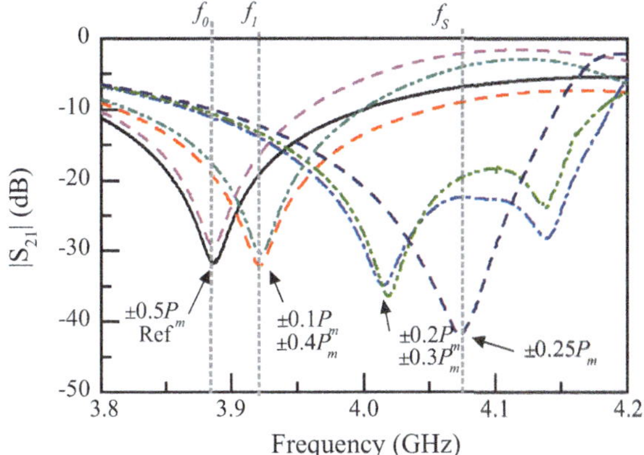

Fig. 4.8 Electromagnetic simulation, using *Keysight Momentum*, of the transmission coefficient for different relative displacements between the stator and the rotor for the rotor with two SRR chains. As compared to Fig. 4.5, except for the curve corresponding to $P_m/4$ displacement, two slightly different responses result for the different displacements between the rotor and stator. The reason is that the two chains of SRRs are not identical. Reprinted with permission from [2]; copyright 2017 IEEE

Fig. 4.9 Attenuation as a function of the rotor displacement at the indicated frequencies for the rotor with two SRR chains. Reprinted with permission from [2]; copyright 2017 IEEE

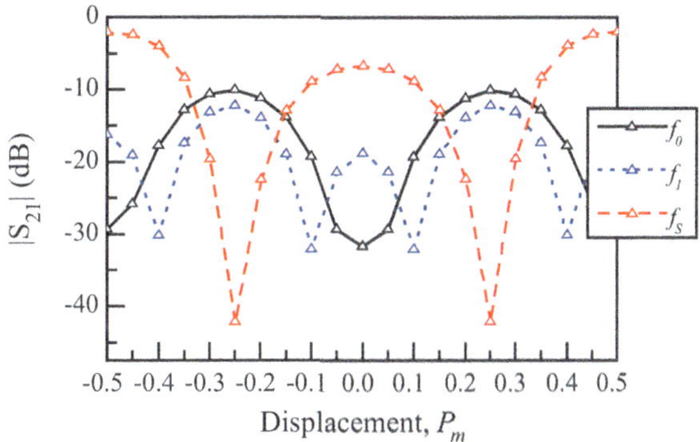

displacement between the stator and the rotor of exactly $\pm P_m/4$ (see the insertion loss for that case in Fig. 4.8), then only two attenuation peaks appear (see Fig. 4.9). For this singular misalignment ($\pm P_m/4$, see Fig. 4.10), the two SRR chains exhibit an undistinguishable relative displacement with regard to the corresponding SRR-pair of the stator. This is the reason that explains the presence of only two attenuation peaks in this case. However, note that the depth of the attenuation peaks is stronger since both chains act simultaneously. This effect is also visible in Fig. 4.8, where the resonance peak for this misalignment is clearly more pronounced. Let us call f_s the resonance frequency for this latter case. If the feeding signal is either tuned to f_0 or f_s, two pulses per period appear; otherwise, four attenuation peaks (or pulses) arise.

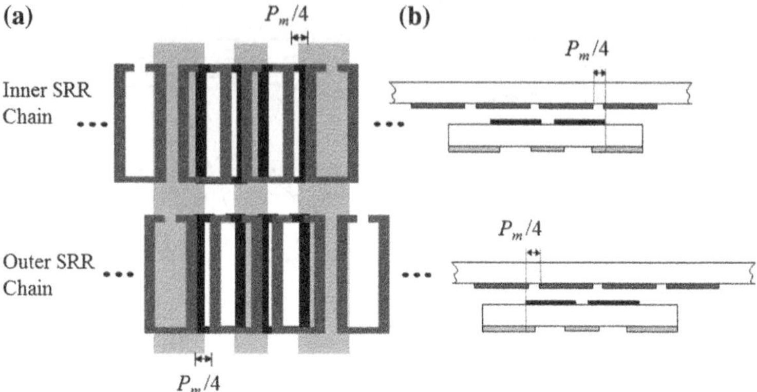

Fig. 4.10 Top (**a**) and cross section (**b**) view of the stator/rotor configuration for the particular case of $\pm P_m/4$ misalignment

Nevertheless, depending on the specific feeding frequency, i.e., if it is close to f_0 or f_s, it is possible that the four peaks are not perfectly visible.

4.3.1.3 Effects of the Air Gap Variation

Let us next discuss the effects of the air gap in more detail. Since the resonances that appear in Figs. 4.5 and 4.8 are due to the combined effect of the SRRs of the stator and rotor, it follows that such resonances should be influenced by the air gap separation between them. Figure 4.11 presents the insertion loss of the CPW for different relative displacements between the stator and the rotor parameterized by the air gap separation. The figure corresponds to the case of a single chain of SRRs for simplicity. As the air gap increases, the frequency range where resonances appear (for different relative displacements) progressively decreases. Note, however, that this constriction is due to the increase of the BC-SRR resonance frequency (the smallest one). The resonance frequency corresponding to a semi-period displacement ($\pm 0.5P_m$) does not experience a significant variation. The reason is that the resonance frequency for such strong misalignment is mainly dictated by the pair of rings of the stator (i.e., the coupling with the rings of the rotor is negligible), and hence the air gap distance has insignificant effect. This narrowing effect is shown in Fig. 4.12, where the limits of the range of resonances as a function of the air gap distance is represented.

The constriction of the resonance frequency range as the air gap increases has two negative effects. On one hand, sensor tuning (i.e., the determination of the feeding signal frequency) is more complex; on the other hand, the discrimination of the two attenuation peaks (provided the frequency of the feeding signal is different from the one of the BC-SRR) is degraded (i.e., the attenuation peaks tend to merge). It is important to guarantee that the air gap separation does not experience significant

Fig. 4.11 Electromagnetic simulation, using *Keysight Momentum*, of the transmission coefficient for different relative displacements (from 0 mm to 1 mm in steps of 0.1 mm) between the stator and the rotor, parameterized by the air gap separation. The graphs correspond to the case of a single chain of SRRs in the rotor. Note that such relative displacements correspond to variation between 0 and $P_m/2$ in steps of $0.05P_m$. The curves corresponding to $P_m/2$ are indicated by dashed lines. Reprinted with permission from [2]; copyright 2017 IEEE

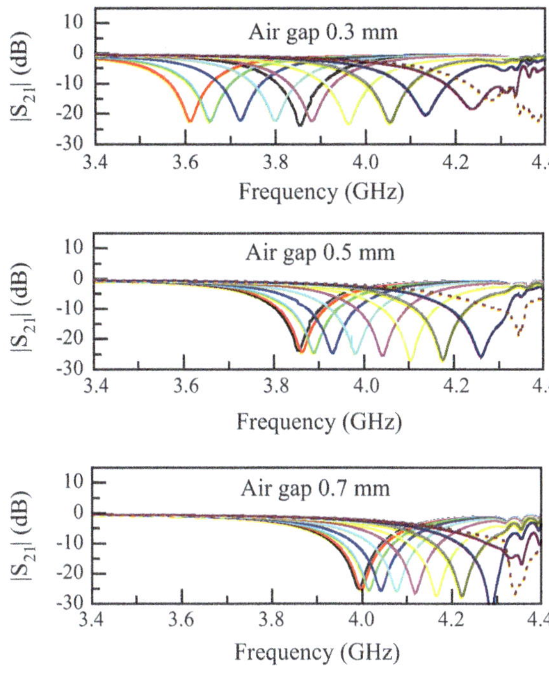

Fig. 4.12 Upper and lower limit of the frequency range where resonances are present, as a function of the air gap distance. Reprinted with permission from [2]; copyright 2017 IEEE

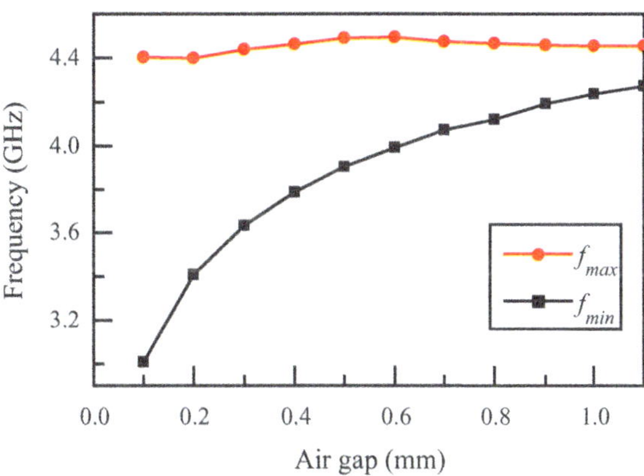

variations during a cycle. If this is not the case, the effect will be a variation of the modulation index with time (see Fig. 4.13). Moreover, note that if the frequency of the feeding signal is set to the one of the perfectly aligned BC-SRR with the smallest air gap, there will not be attenuation if the gap separation increases.

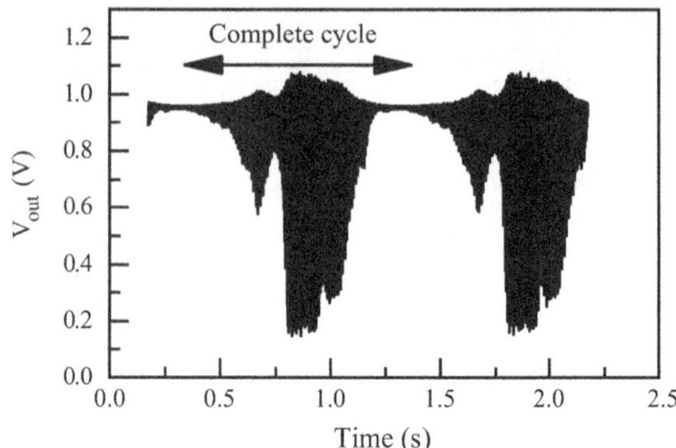

Fig. 4.13 Measured envelope function over a complete cycle with varying air gap separation. Reprinted with permission from [2]; copyright 2017 IEEE

4.3.2 Systems Based on Stator Implemented in Microstrip Technology

In this section, a different approach for the implementation of microwave rotary encoders is discussed. The difference concerns the stator, where a microstrip line loaded with a SRR is used [11]. The system is very similar to the chipless-RFID system reported in Sect. 2.3.2. However, both the rotor and the stator have been re-designed, as compared to the tag and the reader of Fig. 2.28, in order to optimize the angular resolution. Thus, in the stator, the microstrip line is loaded with a rectangular shaped SRR, as shown in Fig. 4.14a. The SRRs of the rotor are identical to those of the stator (see Fig. 4.14b), but oppositely oriented in order to favor the coupling between the SRR of the stator and those of the rotor, as it was discussed in Sect. 2.2. The frequency response of this SRR-loaded line with the presence of the rotor chain on top of it depends on the relative position between the SRR of the line and the SRRs of the rotor. Figure 4.15 depicts several responses corresponding to different displacements between the rotor and the stator inferred from electromagnetic simulation using *Keysight Momentum*. The frequencies designated as $f_{BC\text{-}SRR}$ and f_{stator}

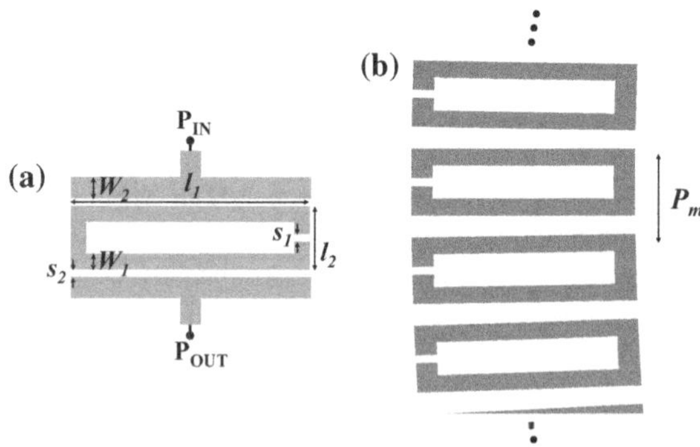

Fig. 4.14 Layout of the stator (**a**) and rotor (**b**). Dimensions are (in mm): $l_1 = 6.2$, $l_2 = 1.6$, $s_1 = 0.2$, $s_2 = 0.2$, $W_1 = 0.4$, $W_2 = 0.535$ and $P_m = 2.0$

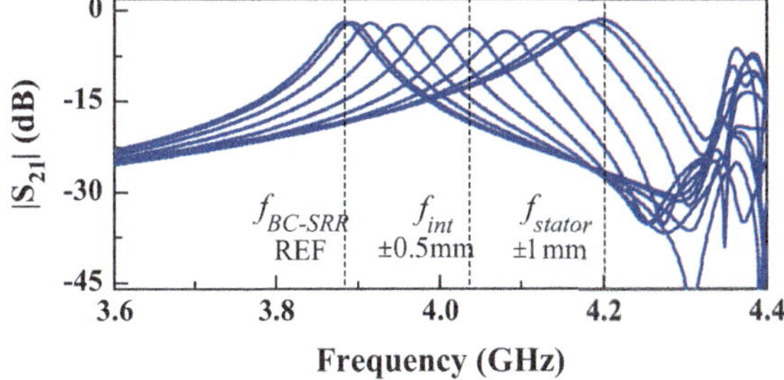

Fig. 4.15 Simulated frequency responses corresponding to different relative displacements between the stator and the rotor. The considered substrate for the stator is the *Rogers RO3010* with dielectric constant $\varepsilon_r = 10.2$ and thickness $h = 0.635$ mm. The considered dielectric constant and thickness for the rotor are $\varepsilon_r = 10.2$ and $h = 1.27$ mm, respectively. The distance (air gap) between the stator and rotor is 0.5 mm. Reprinted with permission from [11]; copyright 2018 IEEE

Fig. 4.16 Variation of the transmission coefficient with the displacement for the indicated frequencies, for an air gap of 0.5 mm. These results have been inferred by electromagnetic simulation. Reprinted with permission from [11]; copyright 2018 IEEE

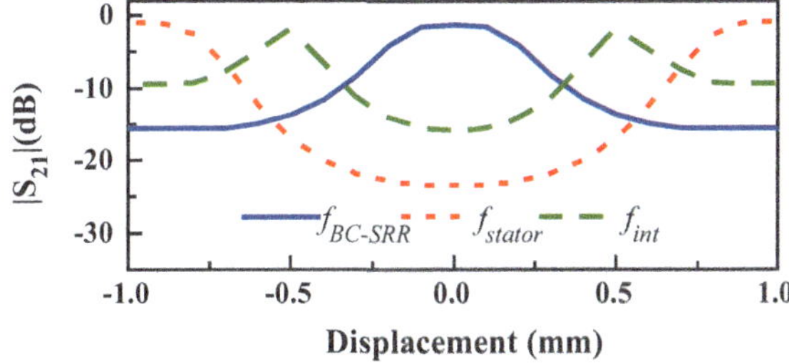

are the frequencies of maximum transmission for the perfectly aligned SRRs of the rotor and stator (i.e., REF position, giving rise to the BC-SRR particles) and for maximally misaligned SRR of the rotor and stator (corresponding to half a period displacement from the REF position), respectively.

Figure 4.16 depicts the variation of the transmission coefficient with the relative displacement between the stator and rotor for three different frequencies, i.e., the extreme frequencies of Fig. 4.15, designated as $f_{BC\text{-}SRR}$ and f_{stator}, and an intermediate frequency, f_{int}. It can be seen that, in all the cases, significant variation of the transmission coefficient arises. The presence of two peaks for intermediate frequencies means that, in general, two pulses per resonant element are expected, unless the carrier signal is tuned to one of the extreme frequencies, $f_{BC\text{-}SRR}$ or f_{stator}.

By varying the air gap distance, the frequency responses for the different relative displacements between stator and rotor experience an upwards (air gap increase) or downwards (air gap decrease) shift. Consequently, the variations of the transmission coefficient with the relative displacement for the frequencies indicated in Fig. 4.16 are different than those shown in that figure. The new results, by considering an air gap of 0.3 mm and 0.7 mm are depicted in Fig. 4.17. It can be appreciated that significant excursions of the transmission coefficient with rotor displacement are obtained in all the cases, which means that the proposed structure is robust against

Fig. 4.17 Variation of the transmission coefficient with the displacement for the same frequencies considered in Fig. 4.16, and air gaps of 0.3 mm (**a**) and 0.7 mm (**b**). These results have been inferred by electromagnetic simulation. Reprinted with permission from [11]; copyright 2018 IEEE

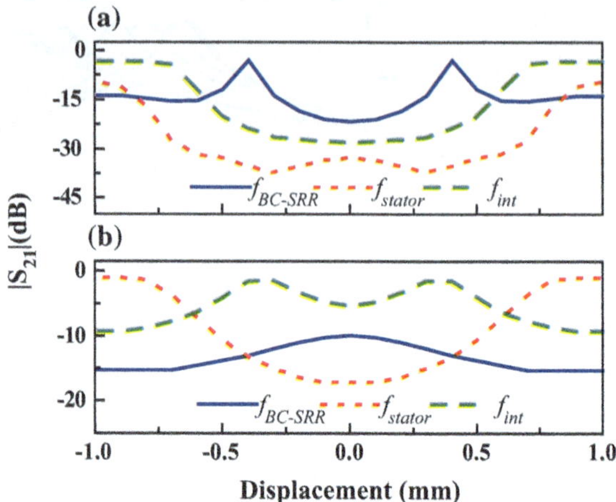

air gap variations, an important aspect due to the difficulty to maintain invariable such distance.

4.3.3   Comparative Analysis

By implementing the stators in microstrip technology, backside isolation is achieved due to the presence of the metallic ground plane in the back substrate side. Such backside isolation constitutes an advantage of the rotary encoders of Sect. 4.3.2 over those based on CPW stators, discussed in Sect. 4.3.1. Although in some circumstances backside isolation may not be a critical aspect, in general, the absence of microwave circuitry (or other interfering sources) in the backside region of the stator cannot be guaranteed. Hence, depending on system requirements, it should be convenient to use the rotary encoders based on SRR-loaded microstrip-line stators. Another advantage of these latter encoders concerns the major robustness against variations in the air gap distance, e.g., caused by misalignments, rotor precession or rotor vibration. This aspect can be appreciated to the light of Figs. 4.11, 4.16, and 4.17. According to Fig. 4.11, relative to stators implemented in CPW technology, for an air gap of 0.7 mm, the harmonic feeding signal of the stator must be tuned within a limited range of frequencies in order to obtain an AM signal at the output port (seen as a train of pulses in the envelope function). Figure 4.12 clearly reveals how such range of usable frequencies is reduced with the air gap. However, in view of Figs. 4.16 and 4.17, corresponding to a system with stator implemented in microstrip technology, it can be seen that the transmission coefficient varies even for gaps of 0.7 mm, regardless of the considered frequency within the range comprised between $f_{BC\text{-}SRR}$ and f_{stator} (see Fig. 4.15).

Note that in the microstrip system presented in Sect. 4.3.2, the presence of a pair of circular chains in the rotor has not been considered. By contrast, it was demonstrated in Sect. 4.3.1.2 that by using two chains of SRRs in the rotor, the *PPR* of the CPW rotary encoder could be enhanced by a factor of two. In the microstrip rotary encoders, the system operates in bandpass configuration, and the solution to double the *PPR* is not as simple as the one reported in Sect. 4.3.1.2, where adding a pair of SRRs in the stator suffices. However, it should be also possible to double the number of pulses in microstrip encoders. A potential strategy (not yet experimentally verified) should consist of using a stator with a pair of parallel-connected SRR-loaded microstrip lines in bandpass configuration. By ensuring that when one SRR of the rotor is perfectly aligned with one of the SRRs of the stator, the other SRR of the stator is completely misaligned with the SRRs of the rotor, two pulses per resonant element of the rotor can be potentially generated.

4.4 Determining the Motion Direction

In the previous sections, the determination of the motion direction has not been considered. One possibility for that purpose is to provide a code to the rotor by detuning certain resonant elements of the chain, or simply by eliminating some of them. From the coding sequence, it can be easily discerned the motion direction of the rotor, i.e., clockwise or counter-clockwise. However, this approach degrades the angular resolution provided the number of pulses, related to the number of functional resonators, is reduced. Note also that eliminating functional resonators in the rotor chain limits the capability to measure instantaneous angular velocities.

An alternative approach to infer the motion direction is to print (or etch) the code in an additional chain, concentric to the chain, or chains, used to determine the angular displacement and velocity [12]. Let us designate such additional resonator chain as rotation direction chain, to distinguish it from the angular velocity chain (or chains), providing the rotation speed. The angular velocity chain (or double-chain) is periodic, with identical SRRs separated the same (tiny) distance, as it was shown in Sect. 4.3. To detect the rotation direction, the corresponding chain should contain certain code, i.e., it should not be periodic. Particularly, by increasingly or decreasingly separating the resonators along the whole chain circumference as the angle varies clockwise, for instance, it is possible to differentiate the motion direction. However, the resonators of the rotation direction chain must be tuned to a significantly different frequency than the one of the resonant elements of the angular velocity chain.

The idea (principle) to separately determine the angular velocity and motion direction is illustrated in Fig. 4.18 [12]. A double-tuned signal, with frequencies $f_{c,v}$ and $f_{c,d}$, should be injected to the input port of the transmission line acting as sensitive part of the stator (let us consider, without loss of generality, a SRR-loaded CPW transmission line). The sub-index c refers to carrier frequency, whereas the sub-indexes v and d are used to differentiate the carrier frequencies corresponding to the velocity and direction chains, respectively. Since $f_{c,v}$ and $f_{c,d}$ must be tuned to the resonance

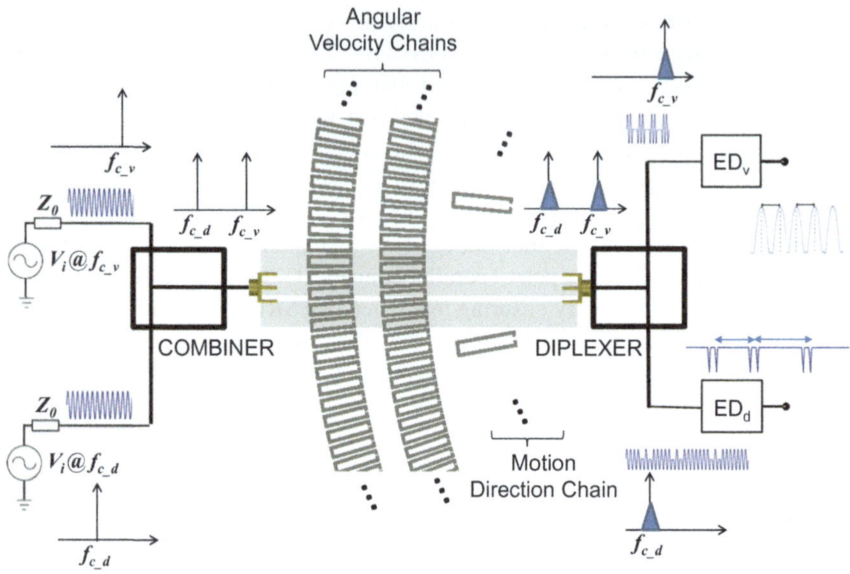

Fig. 4.18 Illustration of the working principle for the independent measurement of the angular velocity and rotation direction. The two carrier signals, tuned to $f_{c,v}$ and $f_{c,d}$, can be injected to the CPW by means of a combiner. Reprinted with permission from [12]; copyright 2018 IEEE

frequencies of the SRRs of the velocity and direction chains, respectively (or close to such frequencies), it follows that the SRRs of both chains must exhibit different resonance frequencies, as mentioned before. Such frequency multiplexing is necessary to be able to separately obtain the information relative to the rotation speed from the information relative to the motion direction. The carrier signals must exhibit significantly different frequencies, $f_{c,v}$ and $f_{c,d}$, in order to avoid overlapping between the sidebands generated by rotor motion at the output port of the transmission line.

By rotor motion, each chain of SRRs (velocity chain and direction chain) modulates independently the amplitude of each carrier signal, providing at the output of the transmission line the spectrum indicated in Fig. 4.18. The information relative to the angular velocity and motion direction is contained in the sidebands of the carrier frequencies $f_{c,v}$ and $f_{c,d}$, respectively. Thus, by means of a diplexer, it is possible to separate each carrier signal (with the corresponding sidebands), resulting in two AM modulated signals in time domain at the two output ports of the diplexer. By means of envelope detectors (ED), the signals containing the relevant information can be extracted, as depicted in Fig. 4.18, and as discussed exhaustively in the previous sections and chapters. The angular velocity is inferred from the distance between adjacent pulses of the envelope function generated at the output port of the envelope detector designated as ED_v. At the output port of the envelope detector designated as ED_d, where the input signal comes from the diplexer channel corresponding to the frequency $f_{c,d}$, the envelope function should exhibit an increasing or decreasing time distance between consecutive pulses, depending on the rotor motion direction. The

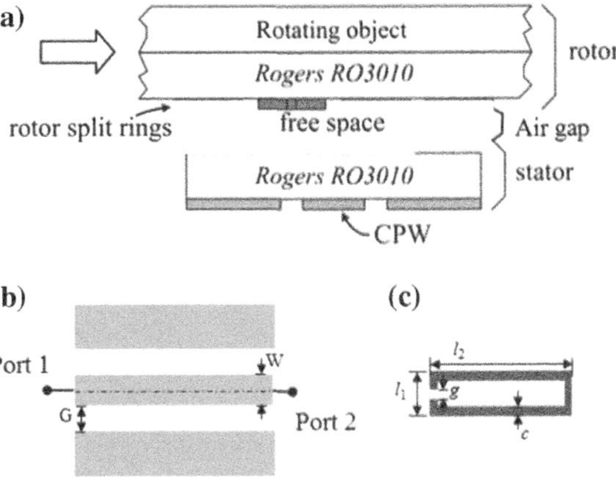

Fig. 4.19 Cross section of the motion direction detector (**a**), and layout of the (**b**) stator and (**c**) rotor. Dimensions are (in mm): $W = 1.3$, $G = 0.9$, $l_1 = 2.0$, $l_2 = 8.4$, $c = 0.4$, and $g = 0.5$. The substrates have relative permittivity of 10.2, thickness of 0.635 mm (stator) and 1.27 mm (rotor), and loss tangent of 0.0023

reason is the varying distance between adjacent SRRs of the direction chain along the rotor circumference, as indicated before. Therefore, the rotation direction can be easily inferred with this strategy.

One important difference between the angular velocity chain and the motion direction chain is the fact that in the latter there is no need to closely space the resonant elements. Thus, coupling between adjacent SRRs, which justifies the presence of SRRs in the stator for the measurement of angular velocities [1, 2], is not an issue for the determination of the motion direction. Consequently, a solution based on the use of unequally spaced single SRRs in the rotor (rather than SRR pairs), without resonant elements in the stator, was proposed in [12] for the sensor used to detect the motion direction, see Fig. 4.19 (additional reasons to justify the absence of resonant elements in the stator can be found in [12]).

As it can be appreciated in Fig. 4.20, as a single SRR crosses the CPW (stator), there are two positions with maximum attenuation. In such positions, the center of the SRR is exactly in the central position of the CPW slots. Moreover, when the SRR is perfectly aligned with the axis of the CPW transmission line, there is total transmission due to the lack of electromagnetic coupling between the line and the resonators. The reason is the perfect cancellation of both the electric and magnetic field lines generated by the CPW in the SRR region [13], [14]. With this configuration, i.e., without SRRs in the stator, the air gap separation is not critical. A variation of the gap distance modifies the notch depth, but such variation does not significantly alter the notch frequency, except for very small gaps (Fig. 4.21). This is due to the fact that the resonance frequency is not related to the broadside coupling between a pair of vertically aligned resonators. Nevertheless, when the air gap is very small, certain variation in notch frequency can be appreciated since the inductance of the SRR may be affected by the metal layer of the CPW. Moreover, with this configuration, the coupling between the line and the SRR is both magnetic (the dominant mechanism) and electric (i.e. mixed coupling [15]), and consequently the notch frequency is more sensitive to the proximity to the CPW metal layer through the capacitive effect

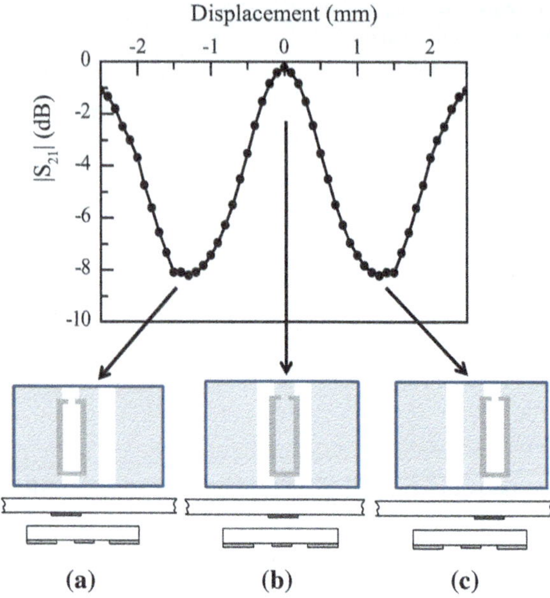

Fig. 4.20 Attenuation as a function of the rotor displacement with simple resonator in the rotor, for the frequency corresponding to the notch in positions (a) or (c). Reprinted with permission from [12]; copyright 2018 IEEE

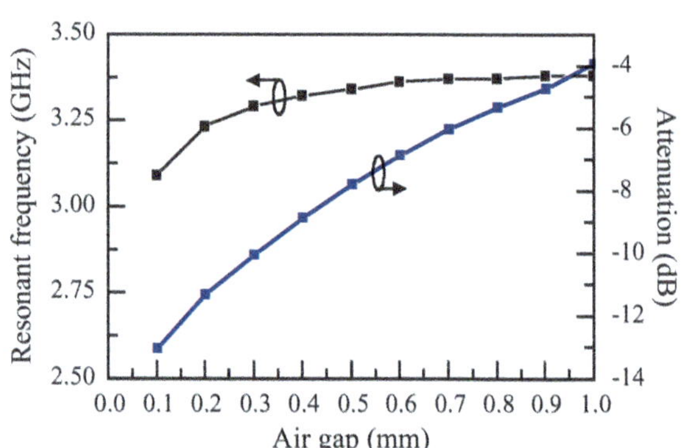

Fig. 4.21 Resonant frequency and maximum attenuation, as a function of the air gap distance for a simple resonator in the rotor when the SRR is in position (a) or (c) in Fig. 4.20 . Reprinted with permission from [12]; copyright 2018 IEEE

between the line and the resonator. With the configuration of Fig. 4.19, when the rotor moves above the stator, the coupling between the line and the resonator is modulated, and consequently the notch depth is modulated as well (Fig. 4.22). It should be taken into account that the dynamic range (maximum attenuation in Fig. 4.22) is relatively small, and therefore it is important to properly adjust the carrier frequency, $f_{c,d}$, to the one of maximum attenuation, or very close to it.

Fig. 4.22 Electromagnetic simulation of the transmission coefficient for different relative displacements between the stator and the rotor for simple SRR in the rotor. **a** displacement of the SRR from the center (axis) of the CPW (where coupling is prevented) to the center of one of the slots (in steps of 0.22 mm); **b** displacement of the SRR from the center of one of the slots towards the CPW ground plane (in steps of 0.44 mm). Air gap of 0.5 mm. Reprinted with permission from [12]; copyright 2018 IEEE

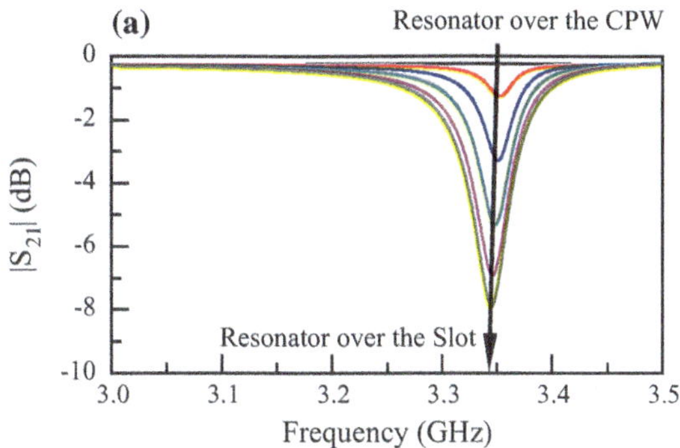

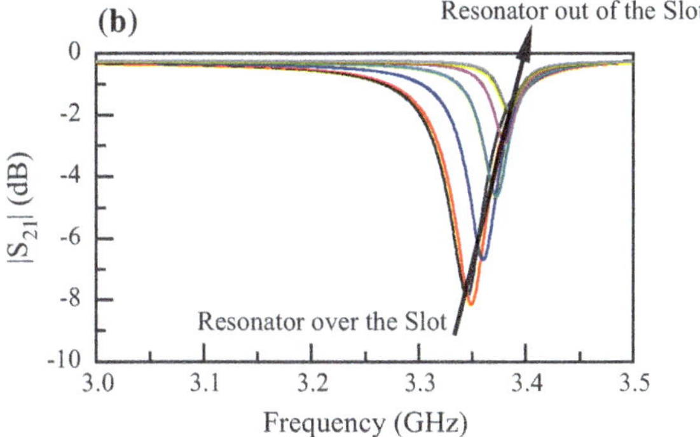

4.5 System Validation

In this section, several prototype examples of microwave rotary encoders, based on the previous stator-rotor schemes, are reported.

4.5.1 *Prototypes with Stators Implemented in CPW Technology*

The first prototype of microwave rotary encoder based on a stator implemented in CPW technology was reported in [1]. In that work, a single SRR chain in the rotor was considered. Then, in [2], it was demonstrated that two pulses per resonant element of the rotor can be achieved, and two prototype devices were presented. In the first angular displacement and velocity sensor, a rotor with a single resonator's chain composed of 300 SRRs was considered. In the second prototype, with a double resonator's chain in the rotor and 600 SRRs in total, it was shown that 1200 pulses per revolution (i.e., *PPR* = 1.200) are achieved by properly tuning the frequency of the feeding harmonic signal.

(a) **(b)**

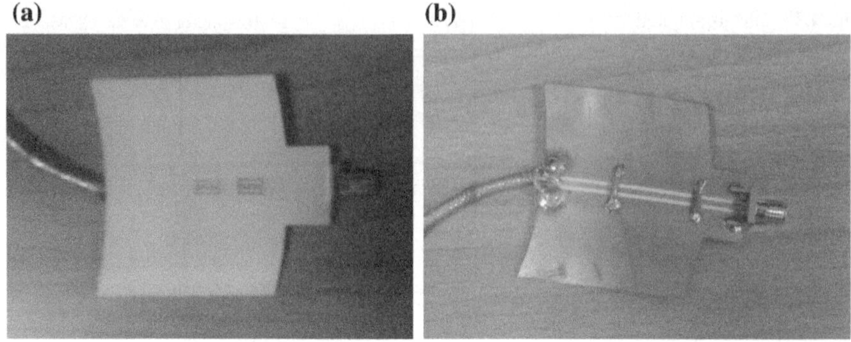

Fig. 4.23 Fabricated stator: **a** resonators, bottom; **b** CPW, top. SRR dimensions (in mm) are those given in [12], namely, $d = 0.4$, $l_1 = 1.6$, $l_2 = 6.2$, $c = 0.4$, and $g = 0.2$

In this book, the validation of the microwave rotary encoders with CPW-based stator is carried out by presenting the characterization results of the system reported in [12], where a rotor with a double-SRR chain to determine the angular displacement and velocity, and with an additional non-periodic SRR chain to discriminate the motion direction, was considered (see Sect. 4.4). The stator transmission line is the one considered in [2], i.e., a 50-Ω CPW implemented on the *Rogers RO3010* substrate with thickness $h = 0.635$ mm and dielectric constant $\varepsilon_r = 10.2$. The photographs (top and bottom views) of the designed stator are shown in Fig. 4.23. SRR dimensions were determined in order to obtain resonances around 4 GHz for the velocity chains [2].

The two periodic SRR chains (velocity chains) of the rotor are simply formed by identical equispaced SRRs (and identical to those of the stator). The non-periodic resonator chain (direction chain) is concentrically etched inside the velocity chains, using only single resonators, rather than pairs, with dimensions given in Fig. 4.19. The measured resonance frequency for an air gap around 0.25 mm is 3.1 GHz. The fabricated rotor is shown in Fig. 4.24, where it can be appreciated that the SRRs of the direction chain are not equispaced and are larger than those of the velocity chain. The reason is that the carrier frequency for velocity measurement has been set to a higher frequency than the carrier frequency for motion direction detection. Note that there are points where the sequence of unequally spaced SRRs is reinitiated. By this means, we avoid an excessive separation between resonant elements and hence an excessive time for the detection of the motion direction. The considered substrate for the rotor is the *Rogers RO3010* substrate with thickness $h = 1.27$ mm and dielectric constant $\varepsilon_r = 10.2$.

As it has been discussed, two AM modulated signals, with carrier frequencies at 4.3 GHz and 3.2 GHz, containing the information relative to the angular speed and motion direction, respectively, are present at the output port of the stator (CPW transmission line). In order to correctly discriminate both information signals, a frequency diplexer was designed and fabricated [12]. The diplexer is implemented in CPW technology and uses pairs of SRRs tuned to $f_{c,v}$ or $f_{c,d}$, depending on the

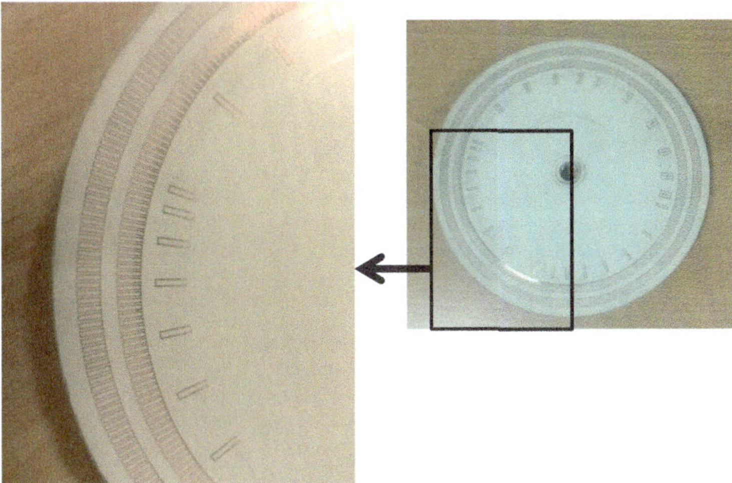

Fig. 4.24 Fabricated rotor with an outer double-SRR chain for velocity measurement and inner unequally spaced SRR chain for direction detection. The radius of the rotor is 101.6 mm. For velocity sensor chains, SRR dimensions are those given in Fig. 4.23. The dimensions of the inner SRRs (detection direction chain) are those given in Fig. 4.19

output channel. Particularly, two SRR stages were used for each diplexer channel. By these means, the isolation between the output ports is enhanced, and the bandwidth for each channel is appropriate for potential adjustments of the operation frequencies in the final implementation. The implemented diplexer and the measured and simulated response are depicted in Figs. 4.25 and 4.26, respectively. The diplexer of Fig. 4.25 can be used inversely as a combiner, in order to inject the harmonic signals with frequencies $f_{c,v}$ and $f_{c,d}$ to the input port of the CPW of the stator (such signals have been generated by means of two function generators).

The previous sensor elements have been assembled with the necessary electronics and instrumentation for measuring and validation purposes. The picture of the complete sensor system is depicted in Fig. 4.27. For the global sensor validation, the CPW transmission line (stator) is fed by the *Agilent E44338C* function generator, providing a harmonic signal tuned to $f_{c,v} = 4.3$ GHz, and by the *Agilent PNA N221A* vector network analyzer, giving $f_{c,d} = 3.2$ GHz (employing the designed combiner). The output port of the CPW is connected to the input port of the designed diplexer through SMA connectors. Each output channel of the diplexer is connected to an envelope detector preceded by an isolator, implemented with the *ATM ATc4-8* circulator for $f_{c,v}$ and with the *ATM ATc2-4* for $f_{c,d}$, in order to protect the CPW from mismatching reflections caused by the diode (a highly nonlinear device). Signal rectification and filtering in each channel, necessary to obtain the envelope function, has been carried out by means of *Avago HSMS-2860* diodes and *N2795A* active probes (with resistance $R = 1$ MΩ and capacitance $C = 1$ pF), connected to an oscilloscope (model *Agilent MSO-X-3104A*). The rotor is attached to a rotating cylinder whose

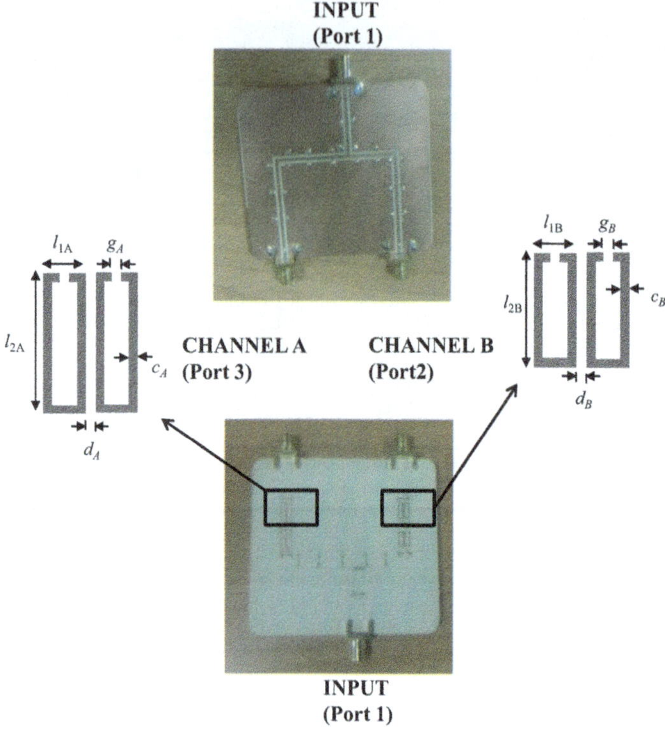

Fig. 4.25 Fabricated diplexer. Dimensions are (in mm): CPW: $W = 1.3$, $G = 0.9$; Resonators: $l_{1A} = l_{1B} = 2.1$, $l_{2A} = 9.1$, $l_{2B} = 6.2$, $c_A = c_B = 0.4$, $d_A = d_B = 0.213$ and $g_A = g_B = 0.2$. The subindexes are used to differentiate between diplexer channel A (providing the angular velocity) and B (providing the motion direction). The distance between the T-junction and the SRR pairs is 30.8 mm and 22.6 mm for channel A and B, respectively (by this means, loading effects are avoided and the isolation between both channels is enhanced). The substrates have relative permittivity of 10.2, thickness of 0.635 mm and loss tangent of 0.0023

Fig. 4.26 Simulated and measured response of the diplexer. Reprinted with permission from [12]; copyright 2018 IEEE

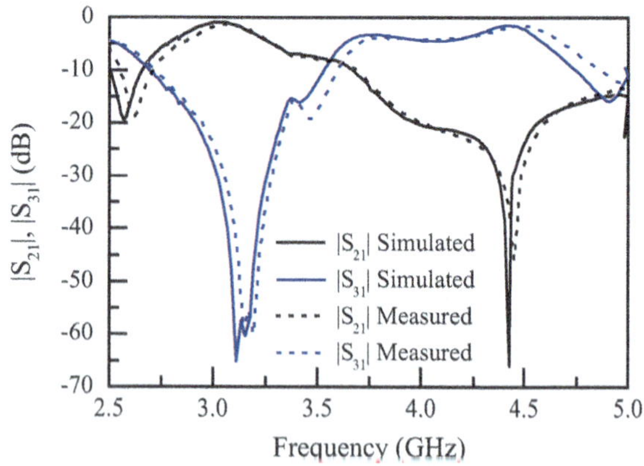

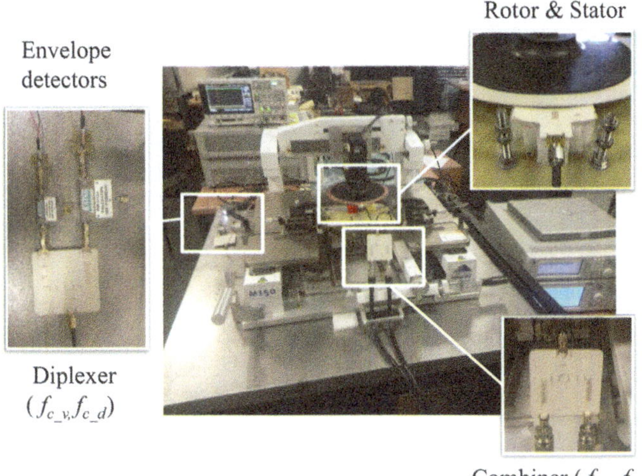

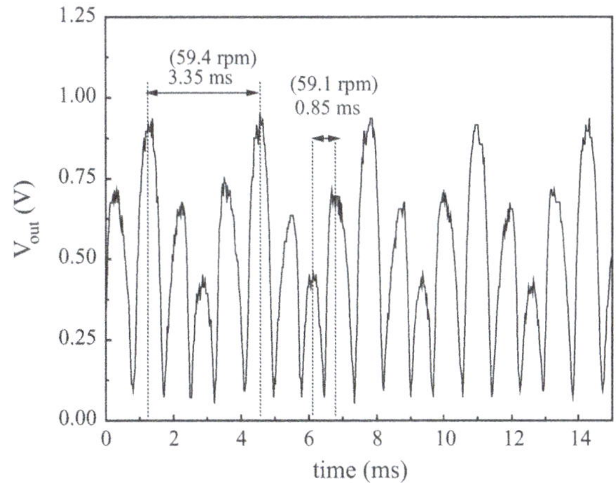

Fig. 4.27 Photograph of the sensor system and the necessary instrumentation

motion (velocity) is controlled by a step motor (model *STM 23Q-3AN*). As compared to first prototypes reported in [1, 2], in the present system [12] the mechanical robustness of the rotor was improved in order to stabilize the air gap (i.e., avoid precession) as much as possible.

To validate the sensor system for angular velocity measurements and for the detection of motion direction, the rotating speed of the step motor was set to 60 rpm. The envelope function at channel A of the diplexer (providing the angular velocity) is depicted in Fig. 4.28, whereas Fig. 4.29 shows the output signal coming from channel B of the diplexer, the one providing the rotation direction. As can be appreciated,

Fig. 4.28 Envelope function of channel A (velocity sensor). The power of the corresponding injected carrier signal, with frequency $f_{c,v}$, is 17 dBm. Reprinted with permission from [12]; copyright 2018 IEEE

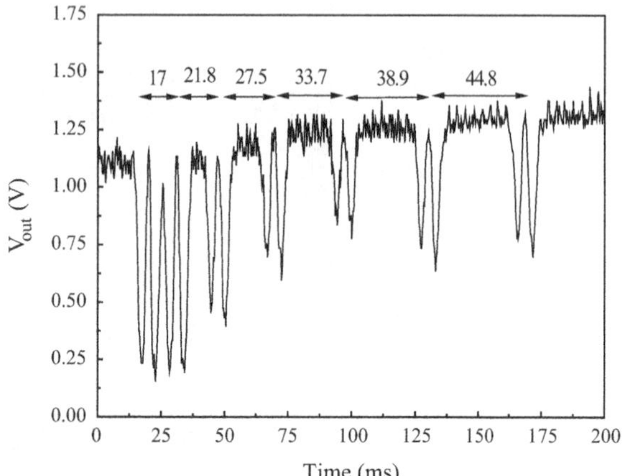

Fig. 4.29 Envelope function of channel B (direction sensor). The power of the corresponding injected carrier signal, with frequency $f_{c,d}$, is 19 dBm. Reprinted with permission from [12]; copyright 2018 IEEE

the distance between adjacent maxima in Fig. 4.28 corresponds to a rotation speed of roughly 60 rpm, in reasonable agreement with the nominal value (1–2% tolerance error is reasonable in the considered step motor). As compared to the results in [2], the periodicity in the measured signal has been improved, due to the abovementioned mechanical improvements, which allow for a major controllability of the air gap between rotor and stator.

On the other hand, the envelope function in Fig. 4.29 shows pairs of pulses increasingly separating with time, which means that rotation is in the direction of increasing inter-resonator distance (direction chain), i.e., clockwise. The average increment, of 6.2 ms, corresponds (taking into account the nominal angular speed of 60 rpm) to a rotation of 2.23°. This value is very close to the theoretical increase in the separation of resonant elements of 2.25°. With these results, the functionality of the sensor is validated, and, particularly, it is demonstrated that the proposed frequency division diplexing is useful to separately infer the angular velocity and motion direction of the rotor.

The automatic determination of the angular velocity and motion direction is possible by means of a post-processing system, consisting of two independent algorithms (a previous digital low-pass filter is employed to eliminate high frequency noise). For the motion direction, the algorithm detects the position of maxima present between two minima. By this means, the distance between two consecutive maxima is determined and then it is checked if such distance increases or decreases. It should be mentioned that there are points where the sequence of SRRs repeats, which means that the increase or decrease in the time distance between adjacent SRRs is not indefinite. Figure 4.30 depicts the flow diagram corresponding to motion direction algorithm.

For the measurement of the angular velocity, once the envelope signal is sampled, such signal is derived. From the zero crosses, corresponding to the maxima and minima of the envelope function, the angular velocity can be inferred. The points of interest for the determination of the angular speed are the maxima. Therefore,

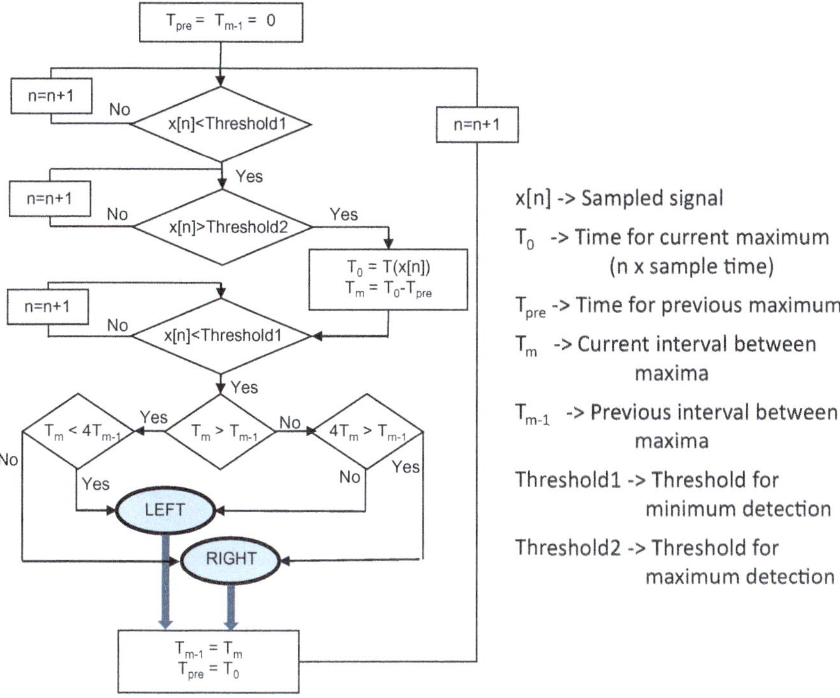

Fig. 4.30 Direction detector signal post-processing flow diagram. Reprinted with permission from [12]; copyright 2018 IEEE

the algorithm detects the time where the derivative of the envelope function changes from positive to negative values. The angular speed is then calculated from the *PPR* of the velocity chain according to

$$\omega_r = \frac{2\pi}{PPR \cdot T_m} \tag{4.1}$$

where T_m, is the time between adjacent maxima, and $PPR = 1200$ in the considered sensor. The flow diagram is depicted in Fig. 4.31.

By using the algorithm of Fig. 4.31 and the sampled data of Fig. 4.28, the angular speed has been automatically determined. A perfect alignment between the rotor and the stator is difficult in practice (particularly the orientation of the stator axis in the radial direction of the rotor). Consequently, the velocity that results from the time difference between each consecutive maxima exhibits certain level of instability (periodic variation around the nominal value). For that reason, it is convenient to average the values, and this has been done by considering four consecutive values of time between maxima, T_m. The resulting velocity as a function of time is depicted in Fig. 4.32, where it can be appreciated that the nominal value (60 rpm) is predicted reasonably well.

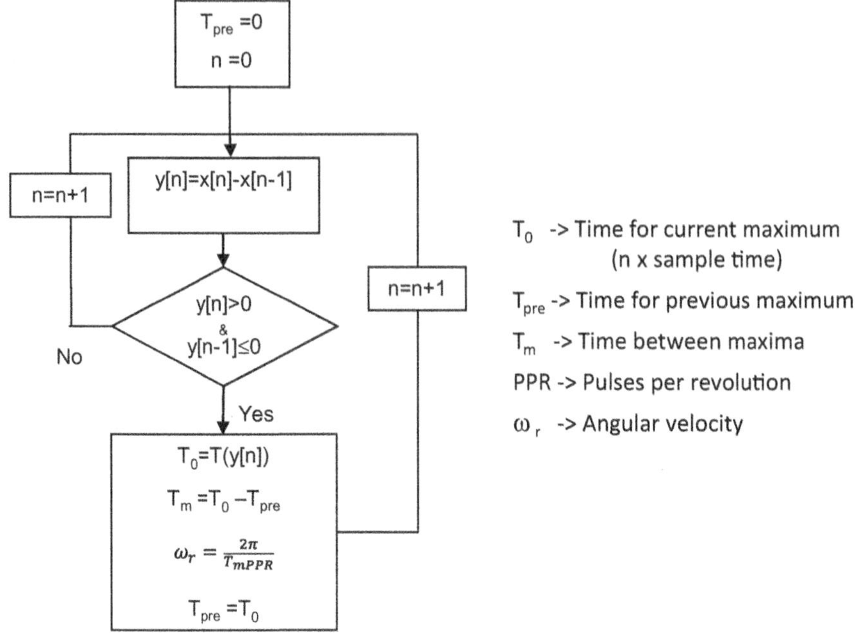

Fig. 4.31 Velocity sensor signal post-processing flow diagram. Reprinted with permission from [12]; copyright 2018 IEEE

Fig. 4.32 Estimated velocity (nominal angular speed is 60 rpm). Reprinted with permission from [12]; copyright 2018 IEEE

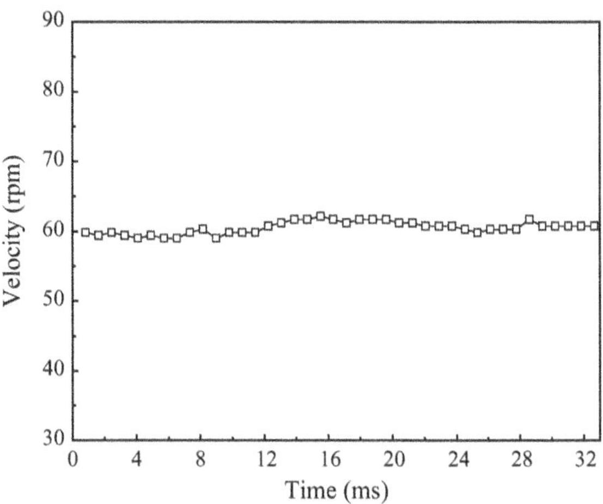

It is worth mentioning that the key factor that limits the maximum measurable velocity in these sensors is the sampling rate. It is reasonable to consider a minimum number of samples per cycle of the envelope function (distance between adjacent pulses). If this minimum number is around 100 samples per cycle (a reasonable value), then it follows that the period (distance between adjacent pulses) should be not less than $100/f_s$ (f_s being the sampling frequency), or equivalently, the maximum rotation frequency should not exceed $f_s/100\,N$, where N is the number of pulses. By considering a (reasonable) sampling rate of $f_s = 100$ MHz, rotation speeds as high as 50.000 rpm can be measured. These velocities are, by far, useful for most applications, including angular velocity measurement in reaction wheels (the application considered in [2]), motion control in industrial applications, servomotors, etc., where angular speeds are typically within the above mentioned limits.

4.5.2 Prototype with Stator Implemented in Microstrip Technology

The photographs of the stator and rotor (with 300 SRRs) corresponding to the microstrip rotary encoder discussed in Sect. 4.3.2 are shown in Fig. 4.33. The experimental setup for measuring the angular velocities is identical to the one shown in Fig. 4.27. The envelope functions corresponding to the three frequencies of Fig. 4.15, for an angular velocity of 60 rpm, are depicted in Fig. 4.34. For the extreme frequencies, $f_{BC\text{-}SRR}$ and f_{stator}, a single pulse per resonant element is achieved. However, for the intermediate frequency, f_{int}, two pulses per resonant element appear, in accordance to the discussion of Sect. 4.3.2. The distances between adjacent (equidistant) minima provide the angular velocities indicated in the figure, in very good agreement with the nominal angular velocity (i.e., the differences are within the typical tolerance errors of the considered stepped motor, around 1–2%).

(a)   (b)

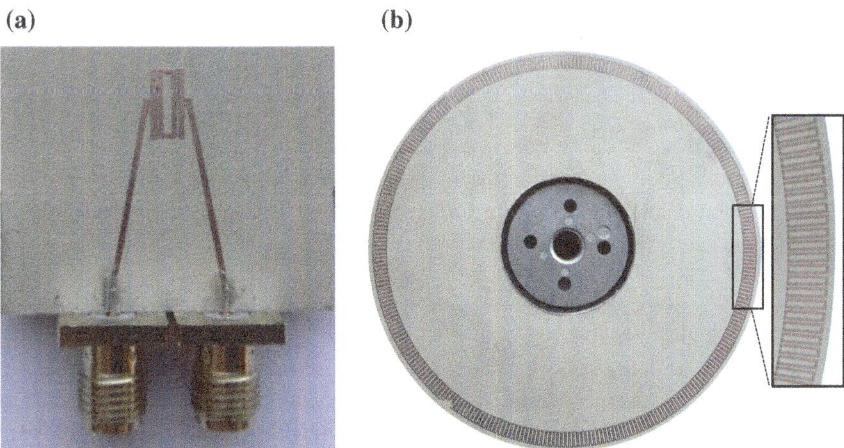

Fig. 4.33 Photograph of the stator (**a**) and rotor (**b**)

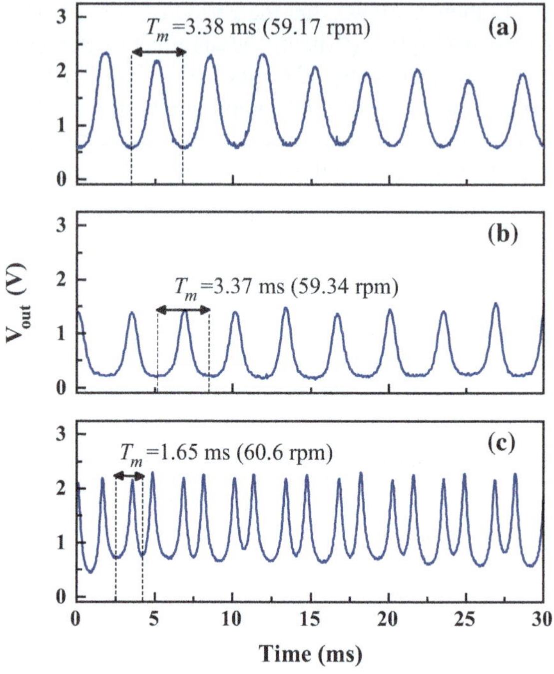

Fig. 4.34 Measured envelope functions for 60 rpm speed and carrier frequencies set to $f_{BC\text{-}SRR} = 3.88$ GHz (**a**), $f_{stator} = 4.20$ GHz (**b**), and $f_{int} = 4.03$ GHz (**c**). The estimated air gap is 0.5 mm. Reprinted with permission from [11]; copyright 2018 IEEE

4.6 Sensor Resolution and Accuracy

Concerning sensor resolution, this aspect should be analyzed for both the measurement of the angular displacement and the measurement of the angular velocity. Angle resolution is given by *PPR* , i.e.,

$$\theta_{res} = \frac{2\pi}{PPR} \tag{4.2}$$

and it is as small as $0.3°$ for the double chain rotor of Fig. 4.24. It should be mentioned that the angular displacement can be inferred from the angular velocity as follows[3]

$$\Delta\theta = \omega_r \Delta t \tag{4.3}$$

but it can be also obtained by simply counting up the number of pulses (in expression 4.3, Δt is the considered time interval).

The resolution in the measurement of the angular velocities is intimately related to time resolution, t_{res}, defined as the minimum time interval that can be resolved. Time resolution is given by the inverse of the sampling frequency. Taking into account that

---

[3]This expression is valid for constant angular velocity. If the rotation speed varies with time, the integral form is needed.

the distance between adjacent pulses is given by

$$T_m = \frac{T_r}{PPR},$$ (4.4)

and that the resolution is t_{res}, the rotation frequency can be expressed as

$$f_r \equiv \frac{1}{T_r} = \frac{1}{(T_m \pm t_{res})PPR}$$ (4.5)

which can be approximated by

$$f_r \approx \frac{1}{PPR \cdot T_m}\left(1 \mp \frac{t_{res}}{T_m}\right)$$ (4.6)

provided $t_{res} << T_m$ (T_r, introduced in expression 4.4, is the rotation period). From (4.6), the resolution in the rotation frequency is

$$f_{r,res} = \frac{t_{res}}{PPR \cdot T_m^2} = PPR\frac{t_{res}}{T_r^2}$$ (4.7)

Angular velocity resolution depends on the number of pulses per revolution and on the rotation speed, as well. For a given rotation speed, the resolution increases (is worst) with PPR. This is due to the fact that T_m decreases. However, by increasing the number of pulses per revolution, variations of instantaneous rotation speed within a revolution can be better detected in time. Note that variations in time lapses smaller than the time between adjacent pulses cannot be detected. If the sampling frequency is, e.g., 1 MHz (the one in the measurements of Fig. 4.28), then the time resolution is $t_{res} = \pm 1\ \mu s$. By considering a rotation speed of 60 rpm (corresponding to $T_r = 1$ s) and $PPR = 1200$, the resolution in the rotation frequency is found to be 1.2 mHz, corresponding to a resolution of 0.072 rpm. Note, that this very good resolution is achieved due to the high sampling frequency. Therefore, the sampling frequency can be relaxed, yet keeping the resolution within good limits. For a significantly higher rotation speed, i.e. 6000 rpm, the resolution is found to be 12 Hz, or 720 rpm. In this case, resolution can be improved by simply increasing the sampling frequency (reducing t_{res}).

The error in the determination of the instantaneous angular velocity depends on the width of the pulses and distance between them. Narrower pulses provide smaller error. Nevertheless, the accuracy can be improved either by averaging over several pulses, or by considering distant pulses. In both cases, however, the penalty is the degradation in the ability of the sensor to detect speed variations in small time intervals. To give a reasonable (very conservative) estimation of the sensor error, if we consider the width of the pulse as given by the value at 98% of the pulse maximum, for the double SRR chain rotor and four pulses per chain period (Fig. 4.28), the ratio between the pulse width and the distance between adjacent pulses is 0.1. This represents a 10% error in time (approximately $\pm$ 5%), but, as mentioned, this value can be reduced by 10 times, i.e., $\pm$ 0.5%, if the considered pulses are separated 10 positions. In terms of error in rotation frequency, these values transform to roughly $\pm$ 5% and $\pm$ 0.5%, respectively.

References

1. Naqui J, Martín F (2016) Application of broadside-coupled split ring resonator (BC-SRR) loaded transmission lines to the design of rotary encoders for space applications. In: IEEE MTT-S International Microwave Symposium (IMS'16), San Francisco, May 2016
2. Mata-Contreras J, Herrojo C, Martín F (2017) Application of split ring resonator (SRR) loaded transmission lines to the design of angular displacement and velocity sensors for space applications. IEEE Trans Microw Theory Techn 65(11): 4450–4460
3. Naqui J, Martín F. (2013) Transmission lines loaded with bisymmetric resonators and their application to angular displacement and velocity sensors. IEEE Trans Microw Theory Tech 61(12): 4700–4713
4. Naqui J, Martín F (2014) Angular displacement and velocity sensors based on electric-LC (ELC) loaded microstrip lines. *IEEE Sens J* 14(4): 939–940
5. Naqui J, Coromina J, Karami-Horestani A, Fumeaux C, Martín F (2015) Angular displacement and velocity sensors based on coplanar waveguides (CPWs) loaded with S-shaped split ring resonator (S-SRR). Sens 15:9628–9650
6. Schurig D, Mock JJ, Smith DR (2006) Electric-field-coupled resonators for negative permittivity metamaterials. Appl Phys Lett 88, paper 041109
7. Horestani A, Abbott D, Fumeaux C (2013) Rotation sensor based on horn-shaped split ring resonator. IEEE Sens J 13(8): 3014–3015
8. Sipal V, Narbudowicz AZ, Ammann MJ (2015) Contactless measurement of angular velocity using circularly polarized antennas. IEEE Sens J 15(6): 3459–3466
9. Marques R, Medina F, Rafii-El-Idrissi R (2002) Role of bi-anisotropy in negative permeability and left handed metamaterials. Phys Rev B 65, paper 144441
10. Marqués R, Martín F, Sorolla M (2008) Metamaterials with negative parameters: theory, design and microwave applications. John Wiley, Hoboken, NJ
11. Mata-Contreras J, Herrojo C, Martín F (2018) Electromagnetic rotary encoders based on split ring resonators (SRR) loaded microstrip lines. In: IEEE MTT-S International Microwave Symposium (IMS'18), Philadelphia, Pennsylvania, June 2018
12. Mata-Contreras J, Herrojo C, Martín F (2018) Detecting the rotation direction in contactless angular velocity sensors implemented with rotors loaded with multiple chains of split ring resonators (SRRs). IEEE Sens J 18: 7055–7065
13. Naqui J, Durán-Sindreu M, Martín F (2011) Novel sensors based on the symmetry properties of split ring resonators (SRRs). Sens 11:7545–7553
14. Martín F, Naqui J, Fernández-Prieto A, Vélez P, Bonache J, Martel J, Medina F (2017) The beauty of symmetry: common-mode rejection filters for high-speed interconnects and balanced microwave circuits. IEEE Microw Mag 18: 42–55
15. Naqui J, Durán-Sindreu M, Martín F (2013) Modeling split ring resonator (SRR) and complementary split ring resonator (CSRR) loaded transmission lines exhibiting cross polarization effects. IEEE Ant Wireless Propag Lett 12:178–181

Chapter 5
Concluding Remarks and Future Prospects

The main contribution of this book concerns the unconventional time-domain approach for the implementation of chipless-RFID systems and related sensors. In such systems, the tags are electromagnetic encoders consisting of chains of metallic inclusions (resonant elements or strips) printed or etched on a dielectric substrate (rigid or flexible, including plastic and paper). The reader is an element able to detect the presence of functional inclusions in the chain, when the tag is displaced at short distance over the sensitive part of the reader. Thus, in this system, tag reading proceeds by proximity (through near-field coupling) between the tag and the reader, in a time-division multiplexing scheme, where the ID code is inferred bit by bit sequentially [1]. Such ID code is given by the envelope function generated by tag motion in a resonator-loaded transmission line (the sensitive part of the reader) fed by a harmonic (interrogation) signal. Thus, contrary to the other time-domain chipless-RFID systems (mostly based on time-domain reflectometry) and to the state-of-the-art frequency-domain and hybrid chipless-RFID approaches, the system reported in this book does not require a wideband interrogation signal. For tag reading, a harmonic (i.e., single tone) interrogation signal suffices (as mentioned above), as far as the metallic inclusions of the tag chain are identical (with the exception that some of them may by absent, or not functional, according to the specific ID code). In other words, the spectral bandwidth occupied by the electromagnetic encoders is virtually null, and this is a fundamental aspect with direct impact in the data storage capacity of the tags. Whereas in frequency-domain or hybrid tags the number of bits is limited by the spectral bandwidth (due to the fact that the resonant elements used for coding purposes are tuned to different frequencies), in the proposed system, based on the so-called time-domain signature barcodes, the number of bits is only limited by the length, or area, of the tag chain. Particularly, as it has been reported in Chap. 3, 80-bit tags implemented on paper substrate, occupying an area of 9.44 cm^2 and a length of 28.2 cm (including the header bits), have been demonstrated to be readable [2]. The corresponding data density per surface (DPS = 8.47 bit/cm^2), or per length (DPL = 2.84 bit/cm), as well as the achieved number of bits, are very competitive, and the cost of the tags is substantially below the price of chipped tags.

© Springer Nature Switzerland AG 2020

F. Martín et al., *Time-Domain Signature Barcodes for Chipless-RFID and Sensing Applications*, Lecture Notes in Electrical Engineering 647,
https://doi.org/10.1007/978-3-030-39726-5_5

In the first reported chipless-RFID systems based on time-domain signature barcodes, the tags were implemented as chains of printed (or etched) rectangular- or square-shaped resonators [3, 4]. In order to optimize the shape factor of the tags, encoders based on linear strips oriented in the direction orthogonal to the axis of the chain were proposed. In Chap. 2 (subsection 2.3.3.3), 100-bit tags based on 6.4 mm $\times$ 0.2 mm strips separated 0.4 mm were presented [5]. In such tags, the per unit length and per unit surface densities are DPL $=$ 12.5 bit/cm and DPS $=$ 26.04 bit/cm^2, respectively, although such tags were implemented on a rigid substrate. The resulting 100-bit tags occupy an area as small as 3.84 cm^2, and the total length of the tag chain is 6 cm, resulting in a very reasonable shape factor taking into account the high number of bits of the tags.

Important aspects related to industrial needs have been discussed in the book. In particular, it has been demonstrated that tag programming and erasing is possible by simply cutting or short-circuiting the metallic inclusions of the tag chain, making them inoperative, or vice versa [6]. By this means, all-identical tags may be massively manufactured (e.g., by means of industrial printing processes, such as rotogravure, off-set or screen-printing), and then programmed in a later stage by means of low-cost processes. For instance, inkjet printing is useful to short-circuit the inclusions, whereas cutting them can be achieved, e.g., by means of laser ablation or through a milling machine. Moreover, it has been demonstrated that the electromagnetic encoders proposed in this book can be implemented in flexible substrates, including plastic substrates [7] and ordinary paper [1], thereby contributing to further reduce the cost of the tags (a key aspect for the application of the chipless-RFID technology in real scenarios).

For the functionality of the reported time-domain chipless-RFID approach, proper alignment and distance between the tag and the reader is required. Nevertheless, the system is tolerant to certain levels of misalignments and distance variations, as it has been discussed in Chap. 2. From the tolerance analysis carried out by considering different implementations of the approach, it can be concluded that the requirements of the mechanical system necessary to guide the tag over the reader are not very stringent. It is also important that the relative velocity between the tag and the reader is maintained as much constant as possible during tag reading. Indeed, such velocity must be well known for tag reading, and for that purpose header bits may be used (the velocity is simply given by the time interval between adjacent peaks, or dips, in the envelope function). An alternative to avoid the need of a uniform and well-known relative velocity between the tag and the reader is synchronous reading [8], also discussed in the book. Obviously, synchronous tag reading can be achieved by considering two parallel chains and two independent readers, where one of the chains, with all-functional metallic inclusions, is used to determine the velocity or to directly dictate the reading instants of the reader used to obtain the ID code. An alternative for synchronous tag reading is to use a single reader based on two sensitive elements (resonators), as it has been reported in Chap. 3. In this case, the resonant elements of the reader are tuned to different frequencies, and consequently a combiner and a diplexer to separate the envelope functions providing the tag velocity and the ID code are required, but this does not represent significant system complexity or cost (note that the priority in cost reduction concerns the tags, rather than the reader).

In Chap. 3, several applications of the reported system have been highlighted and discussed, including secure paper [1, 2], proximity sensors [9], and motion control [10–13]. In this later case, the encoders should be implemented with all-functional inclusions. From the distance between adjacent peaks, or dips, in the envelope function, the relative velocity between the tag (the moving element) and the reader (static element) can be determined, and from the cumulative number of peaks, or dips, the linear displacement is inferred (the spatial resolution is given by the distance between adjacent inclusions). A specific application of the proposed system within the framework of motion control concerns electromagnetic rotary encoders [12, 13] (Chap. 4 has been explicitly dedicated to this topic). In this case, the inclusions (resonators) are arranged circularly, along the perimeter of the rotor, a disc made of a dielectric material. Such encoders constitute an alternative to optical encoders (made of metallic discs with apertures) in applications subjected to harsh environments (e.g., space applications), pollution, dirtiness, etc., where optical encoders may find limitations. Moreover, the lack of optical elements in the proposed electromagnetic encoders may represent also a low-cost approach for the determination of angular displacements and velocities. Although it is not possible to compete against optical encoders in terms of spatial resolution (related to the number of peaks per revolution of the encoder), electromagnetic rotary encoders with up to 1.200 pulses per revolution have been presented in Chap. 4 (pointing out the potential of such encoders in many applications).

However, probably the key application of the proposed electromagnetic encoders concerns secure paper and authentication. It has been demonstrated that the metallic tag inclusions can be directly printed on paper, including not only high quality paper [6], but also ordinary paper [2]. Such metallic patterns provide a unique identification code that can be used for identification purposes of the document in which they are printed, or simply to avoid counterfeiting and unauthorized copies. Obviously, the metallic patterns can be photocopied, but such copied tags cannot be read with the dedicated readers (by contrast, optical barcodes can be easily reproduced–photocopied– and read, and the copies have the same functionality as the original codes). Corporate and official documents, certificates, ballots, banknotes, medical prescriptions, authenticity cards, etc., are some of the potential candidates to be secured and authenticated with the time-domain chipless-RFID tags reported in this book. Reading by proximity is necessary, but this may be beneficial in terms of system confidence against eavesdropping and spying. A specific application of the approach to medical prescriptions is detailed in Chap. 3.

Concerning future prospects, further efforts are required in order to implement synchronous reading systems with improved data density. The preliminary results presented in Chap. 3 demonstrate that the proposed strategy, based on a single reader, is promising, but the reported system exhibits a limited data density per length and per surface.

The implementation of time-domain signature barcodes based on linear metallic strips with unprecedented data density and capacity (see subsection 2.3.3.3) has

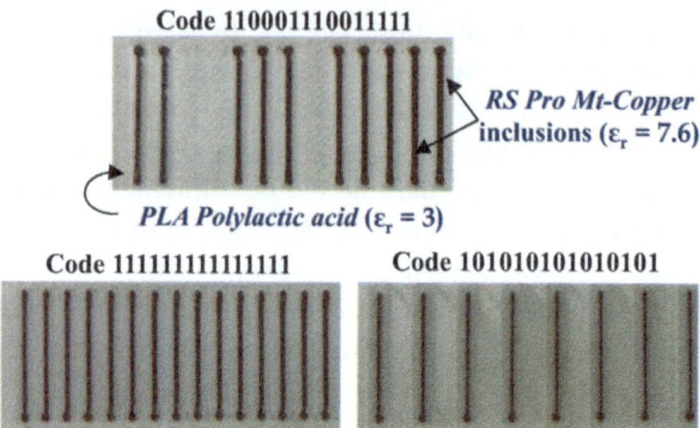

Fig. 5.1 Photograph of 15-bit encoders with the indicated ID codes. The dimensions of the inclusions are length 18 mm and width 0.4 mm, whereas their separation is 3 mm (corresponding to an encoder period of 3.4 mm). The estimated thickness of the 3D-printed substrate is 0.2 mm, and the estimated thickness of the dielectric inclusions is 0.2 mm

been carried out by etching the strips on rigid microwave substrates, and system functionality has been validated. However, tag implementation by printing the strips on ordinary paper has not been reported so far. For that purpose, redesign of the reader and tags might be necessary, in order to guarantee system functionality. Thus, this constitutes a future research activity, where the interest is to optimize the data density with the tags implemented on ordinary paper.

Another topic that can be further developed concerns all-dielectric electromagnetic encoders based on permittivity contrast, recently reported [14] (such encoders can be used either as chipless-RFID tags or as sensors for motion control applications). In these encoders, the main relevant aspect is the lack of metallic elements. Indeed, the tags consist of a dielectric substrate with chains of dielectric inclusions (made of apertures or other high, or low, dielectric constant materials). Lower cost and major robustness against mechanical wearing are potential advantages of these permittivity contrast encoders over the previous electromagnetic encoders based on printed, or etched, metallic elements. The working principle for tag reading of such encoders is very similar to the one of metallic electromagnetic encoders. That is, the presence of dielectric inclusions is detected by proximity, by means of a dedicated reader, consisting of a permittivity sensor, able to detect local changes in the dielectric constant of its surrounding medium. For that purpose, a resonator-loaded line is a good option, as first demonstrated in [14], where a complementary spiral resonator (CSR) loaded microstrip line was used as the sensitive part of the reader. Then, in [15], a microstrip transmission line loaded with a linear slot resonator and a series gap was demonstrated to be a good reader alternative in order to optimize the data density per surface and per length of the tags.

Let us illustrate the potential of permittivity contrast encoders by means of a specific implementation, consisting of 15-bit 3D-printed tags and the corresponding reader [15]. The photograph of three fabricated tags, with the indicated codes, is depicted in Fig. 5.1, whereas Fig. 5.2 depicts the topology of the sensitive part of the

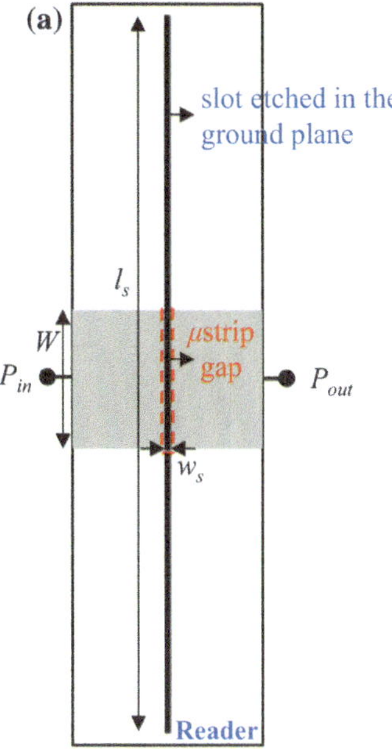

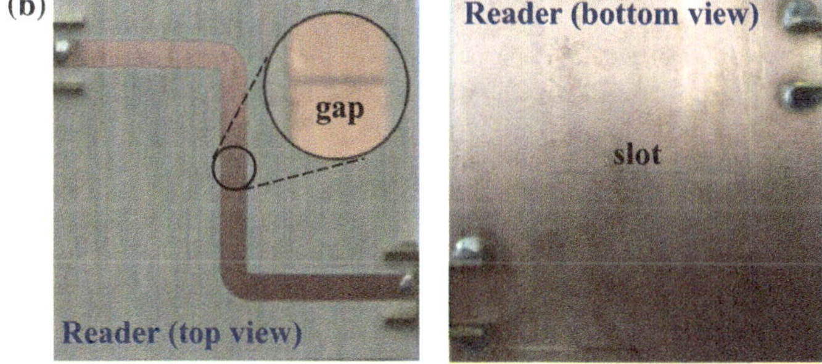

Fig. 5.2 Topology (**a**) and photograph (**b**) of the proposed reader (sensitive part) and relevant dimensions (in mm). $W = 3.98$, $l_s = 24$, $w_s = 0.2$

reader and a photograph of the fabricated device. For tag fabrication, the *Ultimaker 3 Extended* 3D printer was used. Such 3D printer is based on fused filament fabrication, where a continuous filament of a thermoplastic material is used. In the considered encoders, the inclusions are strips of *RS Pro MT-Copper* (with estimated dielectric constant $\varepsilon_r = 7.6$) 3D-printed on top of a dielectric substrate of *PLA Polylactic acid* (with estimated dielectric constant $\varepsilon_r = 3.0$), also 3D-printed. The envelope function of the fabricated permittivity contrast encoders was obtained by means of the fabricated reader and the experimental setup used for reading the previous metallic

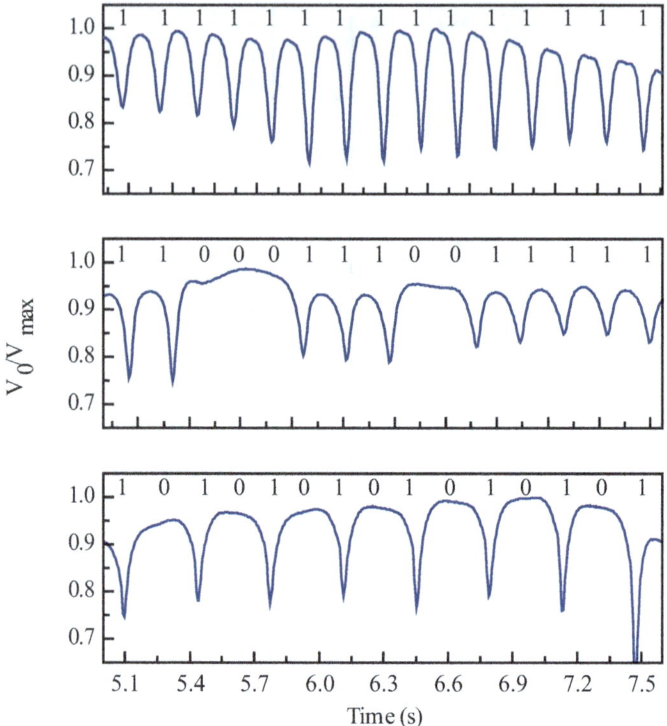

Fig. 5.3 Measured normalized envelope functions for the 3D-printed encoders based on high dielectric constant inclusions of Fig. 5.1. Reprinted with permission from [15]; copyright 2020 IEEE

encoders. The results are depicted in Fig. 5.3, where the presence of inclusions (logic state '1') is revealed by the presence of dips in the envelope function. For tag reading, the encoders are displaced over the reader, with the strip inclusions parallel oriented to the slot resonator of the reader (Fig. 5.4). The presence of inclusions on top of the slot resonator substantially modifies its capacitance (and hence its resonance frequency). The consequence is the modification of the transmission coefficient, and particularly at the frequency of the interrogation signal, thereby modulating the amplitude of such signal at the output port of the reader line (see further details in [15], where the need to include a series gap in the reader line in order to obtain a high modulation index is justified).

With the dimensions of the encoders of Fig. 5.1, the resulting data density per unit surface and per unit length are DPS = 1.66 bit/cm^2 and DPL = 2.99 bit/cm, respectively. These values are not as good as those of the metallic encoders based on linear strips reported in subsection 2.3.3.3. The reason is that in the encoders of Fig. 5.1, the required separation between the dielectric inclusions, for proper discernment of their presence or absence, is 3 mm, resulting in an encoder period of 3.4 mm. This period is not as small as the one achieved in encoders based on chains of transversally oriented

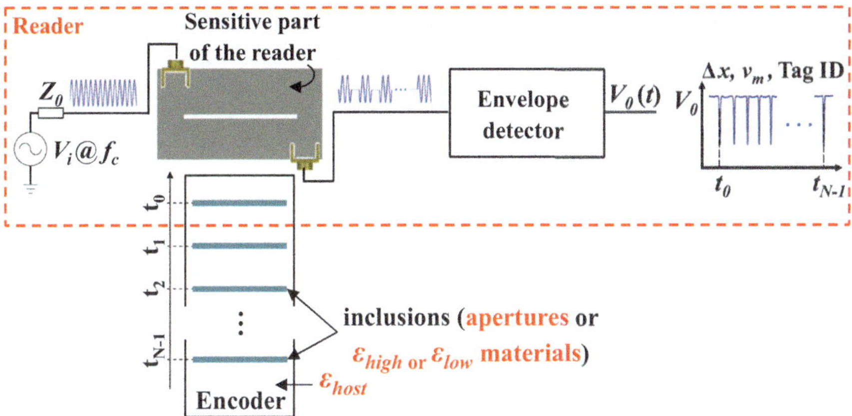

Fig. 5.4 Sketch of the proposed reader-encoder system based on linear dielectric inclusions. The inclusions of the encoder may be high (or low) dielectric constant materials as compared with the dielectric constant of the host substrate, ε_{host}, but such inclusions may be simple apertures

metallic strips, where a period as small as 0.6 mm was achieved (see Chap. 2). The reason is that metals are more efficient than dielectrics in modifying the capacitance of a resonant element through non-contact. Namely, since dielectrics do not "absorb" the electric field lines, as metals do, they need more spatial extension than metals in order to produce the same capacitance (and hence resonance frequency) variation. Although the reported all-dielectric permittivity contrast encoders cannot compete against encoders based on metallic inclusions (etched or printed) in terms of resolution and data density, metallic elements are sensitive to wearing and aging effects, potentially caused by chafing, rubbing, etc. Thus, permittivity contrast encoders constitute a good option in applications where a long-life use is required (at the expense of more limited performance as compared to metallic-based encoders).

It should be mentioned that 3D-printing encoding can be directly carried out during manufacturing of 3D-printed products, e.g., by generating an ID code based on apertures or on dielectric inclusions. This does not represent an extra production cost, and, at the same time, the use of a dedicated tag is avoided. Obviously, the use of optical barcodes is a very low cost alternative (extremely extended nowadays). However, such barcodes may be subjected to wearing, can be manipulated, and their data capacity is limited. By contrast, in the proposed permittivity contrast encoders, the number of bits is only limited by encoder size and encoder implementation in the considered item or product offers very high robustness against mechanical wearing and exposure to harsh environments (e.g., contacting the products with water or liquids does not deteriorate the encoder). Thus, the reported 3D-printing approach for encoder fabrication may be of interest in various scenarios.

Finally, to end this chapter, let us point out that the functionality of time-domain signature barcodes implemented by means of organic conductive inks has been demonstrated [16]. As far as unsustainable metallic conductors are not used during fabrication, such tags are environmentally friendly and fully recyclable. Nevertheless, preliminary results are presented in [16]. Further research activity is required to obtain robust tags based on such organic materials.

References

1. Herrojo C, Paredes F, Mata-Contreras J, Ramon E, Núñez A, Martín F (2019) Time-domain signature barcodes: near-field chipless-RFID systems with high data capacity. IEEE Microw Mag 20(12): 87–101
2. Herrojo C, Mata-Contreras J, Paredes F, Núñez A, Ramon E, Martín F (2018) Very low-cost 80-bit chipless-RFID tags inkjet printed on ordinary paper. Technol 6(2): 52
3. Herrojo C, Mata-Contreras J, Paredes F, Martín F (2017) "Near-field chipless RFID encoders with sequential bit reading and high data capacity", In: IEEE MTT-S International Microwave Symposium (IMS'17), Honolulu, Hawaii, June 2017
4. Herrojo C, Mata-Contreras J, Paredes F, Ferran Martín (2017) Near-field chipless RFID system with high data capacity for security and authentication applications. IEEE Trans Microw Theory Techn 65(12): 5298–5308
5. Herrojo C, Paredes F, Martín F (2019) Double-stub loaded microstrip line reader for very high data density microwave encoders. IEEE Trans Microw Theory Techn 67(9): 3527–3536
6. Herrojo C, Mata-Contreras J, Paredes F, Núñez A, Ramon E, Martín F (2018) Near-field chipless-RFID system with erasable/programmable 40-bit tags inkjet printed on paper substrates. IEEE Microw Wireless Compon Lett 28: 272–274
7. Herrojo C, Mata-Contreras J, Paredes F, Núñez A, Ramón E, Martín F (2018) Near-field chipless-RFID tags with sequential bit reading implemented in plastic substrates. J Magn Magn Mater 459:322–327
8. Paredes F, Herrojo C, Martín F (2019) An approach for synchronous reading of near-field chipless-RFID tags. In: 10th IEEE International Conference on RFID Technology and Applications (IEEE RFID-TA 2019), Pisa, Italy, Sep 2019, pp 25–27
9. Paredes F, Herrojo C, Mata-Contreras J, Martín F (2018) Near-field chipless-RFID sensing and identification system with switching reading. Sens. 18:1148
10. Herrojo C, Mata-Contreras J, Paredes F, Martín F (2017) Microwave encoders for chipless RFID and angular velocity sensors based on S-shaped split ring resonators (S-SRRs). IEEE Sens J 17: 4805–4813
11. Herrojo C, Muela F, Mata-Contreras J, Paredes F, Martín F (2019) High-density microwave encoders for motion control and near-field chipless-RFID. IEEE Sens J 19: 3673–3682
12. Mata-Contreras J, Herrojo C, Martín F (2017) Application of split ring resonator (SRR) loaded transmission lines to the design of angular displacement and velocity sensors for space applications. IEEE Trans Microw Theory Techn 65(11): 4450–4460
13. Mata-Contreras J, Herrojo C, Martín F (2018) Detecting the rotation direction in contactless angular velocity sensors implemented with rotors loaded with multiple chains of split ring resonators (SRRs). IEEE Sens J 18: 7055–7065
14. Herrojo C, Paredes F, Mata-Contreras J, Martín F (2019) All-dielectric electromagnetic encoders based on permittivity contrast for displacement/velocity sensors and chipless-RFID tags. In: IEEE-MTT-S International Microwave Symposium (IMS'19), Boston (MA), USA, June 2019
15. Herrojo C, Paredes F, Martín F 3D-printed high data-density electromagnetic encoders based on permittivity contrast for motion control and chipless-RFID. IEEE Trans. Microw Theory Techn submitted. https://doi.org/10.1109/TMTT.2019.2963176
16. Herrojo C, Paredes F, Escudé R, Ramon E, Martín F (2019) Near-field chipless-RFID system based on tags implemented with organic inks. In: Asia Pacific Microwave Conference (APMC'19), Singapore, Dec 2019

The manufacturer's authorised representative in the EU is Springer
Nature Customer Service Centre GmbH, Europaplatz 3, 69115 Heidelberg,
Germany. If you have any concerns regarding our products, please
contact ProductSafety@springernature.com

Printed and bound by CPI Group (UK) Ltd, Croydon, CR0 4YY

28/11/2025

02007732-0005